STONE AGE CODE

FROM MONKEY BUSINESS TO AI

SHANE NEELEY

Published by Fort Rock Media

FortRockMedia.com

Author's Website: ShaneNeeley.com

Book's Website: StoneAgeCode.com

Author's Social Media: @chimpsarehungry

Library of Congress Control Number: 2021904961

ISBN (paperback):   978-1-7362669-5-3

ISBN (hardcover):   978-1-7362669-6-0

ISBN (ebook):   978-1-7362669-4-6

Audiobook: available on Audible, Apple Books and more.

Cover Design by Toby Way, with images by Shane Neeley.

Interior Design by Melvyn Paulino.

First Printing, 2021, in the United States of America

10 9 8 7 6 5 4 3 2 1

Dedicated to Emma, Abigail, Tina, and in memory of my great Uncle John.

Thank you for purchasing this book! You can enter your email on ShaneNeeley.com to receive a **free digital sample** of the book *AI Art - Poetry: A Style Transfer Photo Anthology with Poems by (human & non-human) Poets.*

Acclaim for Stone Age Code

Real readers:

> "The book is simply brilliant and genuine, so friendly and stimulating!"

> —**EMILIANO BRUNER, PH.D.**, Hominid Paleoneurology Researcher, Centro Nacional de Investigación sobre la Evolución Humana (Spain)

> "A charming, informative, and thought-provoking read."

> —**ADAM CORNFORD,** poet, journalist, and a great-great-grandson of Charles Darwin.

> "My overall impression as a lifelong professor of literature is that this book is engaging, humorous, thought-provoking, creatively written, and artistically inspired."

> —**ALWIN BAUM, PH.D.**, Professor of Literature, California State University (retired)

Fake acclaim for *Stone Age Code*, written by AI:

> "Shane Neeley, data scientist, biologist, and bestselling author of High Frequency and Data Density, answers each and every AI question you've ever asked."

> —**ACCLAIM-WRITING-ROBOT**

> "Book of the year (so far)."

> —**ACCLAIM-WRITING-ROBOT**

> "Read it, laugh at it, and move on."

> —**ACCLAIM-WRITING-ROBOT**

Table of Contents

Introduction: Lab Monkey

Jumping into my 1994 forest green Subaru wagon and heading to the Oregon National Primate Research Center, I had the feeling I'd embarked on what would be a prosperous career of monkey-knee stem-cell research. The center staff received my online order for monkey knees and were waiting for me to accept them. I brought my blue lunch cooler full of ice, picked up the knees, and drove back to my lab with music blasting from my iPod cassette tape adapter. I was ecstatic; this was 2011, and I felt that a future Nobel Prize in medicine was inside that tissue, on ice in my front seat.

Monkey knees are hairy, but I wasn't in it for the hair. I wanted cartilage! It contained the most potent regenerative material: the stem cell. I was imagining my lab would prove to the world we could generate new knees from old ones. Helping fix every poor grandma and every championship-forfeiting draft decision by the Portland Trail Blazers.

The monkeys I was dealing with are contagious for a deadly form of Herpesvirus B that will kill you in two weeks flat after exposure. In 1997, a twenty-two-year-old research assistant died from an infection at a different primate colony. (A "colony?" As if the monkeys claimed any land). After our requisite monkey-use certification, I had my high-school intern do a lot of the scalpel work. If we wore multiple pairs of gloves, there was more protection against the bloody knife poking through, right? This was serious monkey business.

In the orthopedics research lab, the next poor creature I had to cut up was a New Zealand Flemish Giant Rabbit. To feel better about doing this, I would tell myself that these fluffy things were not cute because of their fat necks and their demineralized, bone-matrix-fused spines.

Later, I trotted my cooler to a doctor's office where a fresh human knee was awaiting me. Thankfully, they left the skin and hair on the human, who was alive and well. He was just trading his old knee for a ceramic one. I put his old one in the cooler and hopped on the university's gondola. I was dreading that someone would ask me, "What's for lunch?"

When I got back, a co-worker asked me if I could help him with a rat-eye cutting procedure. After years of animal sacrifice, radioactivity, and inhaled formaldehyde, I started feeling that research wasn't for me. I love animals, they're the reason I became a biologist. How would my childhood hero, Steve Irwin, feel about this work?

One day I felt sick when a warm pile of freshly killed giant rabbits on a metal cart rolled into the operation room. I told a laboratory assistant I would do anything, all her assays, if she did my rabbits. Thank heaven, she accepted my trade. She stayed down in the operating rooms with piles of bunnies and pigs splayed out. I was joyful to be doing her hydroxyproline tests in the chemical fume hood upstairs. Though I couldn't stomach the gory work, I still believed in the research, and this was the experience I needed before graduate school, I told myself.

I kept doing lab work and enjoying the process of discovery. But it was very depressing when I couldn't find the magical stem cell, called a chondroprogenitor. I scoured hundreds of slices of tissue under the microscope, waiting for just one cell to fluoresce bright green, but the images always remained a dark landscape. The mythical, hypothesized cell either didn't exist, or I was doing something wrong. This failed experiment annoys me to this day, as my own banged-up joints could now use some regeneration.

My research assistant salary of $27K was the same as my student loan balance. I made the minimum payments so I could make rent in Portland. I kept at it for three years because I knew this lab job would help me carry out my dream of getting a doctorate in biology. Becoming a scientist had been core to my identity since age ten, when my grandparents subscribed me to Ranger Rick, the kids magazine from the National Wildlife Federation.

While working, I applied to twelve PhD programs, and to my utter astonishment, every school rejected me. I was confused, hurt, and broke from the application fees. The reason for the rejections was that one of my references had written a negative letter of recommendation. It went in a sealed envelope to all twelve schools, telling them something unfavorable, I don't know what, about my character or prospects as a researcher. I ended up finding this out from another reference, who wasn't supposed to tell me. Why in hell this former boss that I had naively trusted would write this instead of no letter, I still don't know. I saw him once grocery shopping and just wanted to chuck an ear of corn at his face. But nagging rumors of his other employees hating him too were enough punishment, I felt.

In a frenzy, I applied to master's programs, omitting the letter of condemnation, and every program accepted me! This was nice, but it was still short of my dreams of a doctorate. An engineering school gave me a contingent acceptance; I had to finish a year's worth of calculus that wasn't on my transcript. I did the master's program, thinking it would look good should I reapply for a doctorate. The math requirements were absent from my transcript because I don't like math and had only done the minimum needed as a biology undergrad. To complete the course, I took night classes at Portland State University after my day job in the lab.

I held three lab assistant positions. They fired me from the first one because I couldn't clone DNA fast enough (I thought this was weird grounds for termination). The second one, orthopedics genetics research that I loved, laid me off after their grant ran out. Now I was on my third—a lab trying to make as many new virus variants as quickly as possible. Here I learned about the Japanese work ethic, and I could not even come close to keeping up with their typical twelve-hour days. Yes, they liked to show up a little late, past 10:00 a.m., but nobody left until after the boss, the sensei, did. From my apartment when I went home, I could see the lab's lights still on later than any others, often until midnight.

Academia was a hectic life, and I felt like I couldn't keep my head above water. Was I not cut out for science? Were the PhD programs right to reject me? Would I just keep lab-hopping with no security long enough to make a real discovery? On top of those confidence blows, I knew I could not become a scientist when I started sabotaging my own experiments. Ugh, science is hard. I wouldn't completely sabotage the experiment, per se, but semi-unconsciously hold the Bunsen burner longer to my negative bacterial control (the one that should *not* work) than to my positive. This is science heresy, immoral, and I committed it under the psychological pressure academia imposes.

One day, they evacuated the building because someone on a floor below me had spilled a neurotoxin, attempted to clean it up themselves, and didn't tell anyone. Was this an academic-induced sheepishness? A researcher afraid to admit a mistake to the principal investigator? They went home feeling sick and later went to the emergency room where they reported what had happened. Someone else in their lab also went home sick. She didn't know why until authorities found her, and she went to the ER.

One day, instead of beakers and bench work, my boss had me do something in Microsoft Excel. Carry values from a spreadsheet to an online protein weight calculator, then put the results in a new column. It was an ordinary task. One part of the less ordinary job of engineering new virus variants to inject into monkeys. Again, poor monkeys.

I had about a thousand Excel rows to do. Something clicked inside my head where I thought, "Hey, isn't this a thing people write code for?" So I sat, and sat, and sat and learned some *Python* basics.

I had no computer science training. I thought a spreadsheet formula was complicated enough; I thought that was coding. Outside of writing HTML for a memorial website when my childhood guinea pig died, I never put down a line of code. But I saw a path forward, and I'd made sure I did a fantastic job at a script to automate the work. Python is a programming language great for scientific tasks. An engineer I knew suggested I teach myself Python over the more commonly used, at the time, *Matlab*, or my boss's language of choice, *Perl*. Many biologists were reluctant to learn to code, or even use databases. Our lab was forced to when gene sequencing data caused some files to surpass Excel's 1 million row limit.

It blew my sensei's mind when I showed him a scripted solution to something he had been doing manually for years. He then let me take half of my time from bench work to write code instead because he saw that I was effective. This gave me the ability to tell my coworkers, Kei and Yasuhiru, to go to hell with their mouse torture, I had code to write!

This was an enormous boost of confidence. I was effective! I still didn't have a Japanese work ethic, which they seemed to understand, but I had a skill they didn't. This earned me a new title. Not just an assistant, but "Shane-san," a student of the art. We could write programs to automate tasks and make everyone more efficient. I co-authored the postdoc's papers because my programs helped their research. My skills progressed far beyond spreadsheet tasks; I built a bioinformatics simulator for the viruses we wanted to inject in monkeys. This helped the project, but not the monkeys, who had hundreds of distinct organ parts harvested from them. They were meticulously dissected so we knew where in their anatomy our DNA-barcoded viruses attached.

In graduate school, I increased my coding experience by working in a commercial lab where we placed iPods in 98.6°F incubators. The iPod cameras were continually taking images of 3D magnetically-levitated clumps of cells as they grew and were treated with a variety of drugs. I would take the steaming hot iPods out and analyze the images with custom software I built. This lab wasn't Japanese, it was the opposite: Brazilians, who also show up past 10:00 a.m., but leave in the early afternoon to play soccer. This work, and everything I learned in the bioengineering program, made a bonafide coder out of a former lab monkey.

{</>}

I had always had an interest in biology and animals. When I was in middle school, I started a fund with spare change in an empty margarine tub labeled "Monkey-Moolah." Somewhere around Google's five-billionth search ever performed was when I typed "how to buy a monkey" and found a website willing to sell me a white-faced capuchin for $1,500. As time went by, my priority for that amount of cash became getting a car instead. It was bittersweet when I cashed in my moolah for the Subaru wagon.

What attracted me to monkeys was their intelligence. Monkey-owning has its downsides, so I've developed a new passion. No, not monkey research. I feel bad about that, but I know they are an ideal model (a simulation) through which to make human disease breakthroughs. We need primates for advanced stages of research, so it would be harmful to break in and release them all, which is something anyone but those involved in PETA would probably agree with. Regardless, the future may look back in horror at our frequency of keeping conscious, emotional creatures in cages, aquariums and zoos. I certainly support the moratorium on great ape research, and I consider whale captivity imprisonment.

For my purposes, I have different models. Now that I've learned how to code, in the comfort of my home, I can make the next best thing to monkeys: *Artificial Intelligence* (AI).

What follows are simply perspectives from me, an outsider-insider. Not a PhD-level biologist, not a computer scientist, not a paleoanthropologist. But a bioengineer, self-taught in machine learning and highly interested in our biological past: I talk about my evolution, our shared evolution, and give brief instruction on how you too can join the AI revolution.

CHAPTER 1

A Greater Ape Approaches

Homo sapiens are pretty cool, as far as animals go! Despite the mass extinctions and impending ecosystem collapses, it can be argued that animal life itself is doing well. The blue whale is the largest animal known to exist in all of evolutionary history.[1] The biggest animal ever is with us right now, and it's an intelligent mammal with a large emotion-processing limbic system! That's cool, but something new is coming. Something, arguably, even cooler than a whale.

The robots are coming ... No, wait, they're here! They're writing this book?! Yes, AIs I've trained wrote swaths of this book, or at least inspired my writing. I solicited help from AIs for the cover art, author bio, and book description as well. I trained the robot to write in a style (mine), and then I read hundreds of pages of its printed output to get ideas. Periodic retraining of the AI while writing this book served as a sort of developmental editor. Topics generated from training on my writing helped break writer's block. Any direct writings from the robot are contained in the designated excerpts.

I often use these concepts interchangeably: *model*, *ML*, *AI*, and *robot*. However, there is a distinction. I think of a model as a specific piece of code responsible for training and predicting. ML (machine learning) is a field of engineering and research. ML encompasses the algorithms (models) that can improve their usefulness through training. AI refers to an ML-based system that provides some real-world utility. An AI will involve one or more models and methods to store, process, and serve up info to a user. Above all, I prefer to just say, "I'm building robots." A robot incorporates the other terms and has human characteristics.

---

[1]   Pinet, Paul R.. Invitation to Oceanography. United States, Jones & Bartlett Learning, LLC, 2019.

Frequently, I try to make connections from computers to biological life. Considered as an organism, a robot's fundamental biochemistry is ML, mathematics, and theory. Its calculus equations are like the Krebs cycle for a cell's metabolism, studded with must-have formulas. The models it runs are akin to genes and proteins; a model has a function, transcribed and expressed through hyperparameters of different cellular organs and epigenetics. AI is the chromosome, a collection of genes. The complete set of chromosomes (the ploidy) directs the organism. A robot culminates in a set of AIs that determine what its life will be. Evolutionary principles shape these robots through time. The data is its ecosystem, it must adapt to the inputs from the environment.

There may be a step beyond this definition of robot that has yet to be invented: a form of true "artificial life"; something that has a theory of mind dependent on the external world and a physical body that feels, moves, learns and changes its environment. We don't have that yet. At least, I don't know how to make it. The robots I consider are ones I can spawn on my computer that can learn and generate language.

With this concept in mind, that AI is like life and can obey natural laws, we can draw a lot of interesting conclusions. To begin, the driving force of selection for the best AIs will be our hands. Though our hands were driven by natural selection, our neuropsychology is full of incredible art and spirituality in addition to rational science. Therefore, AI has a chance to follow suit and become an interesting, emotional organism, like the walking apes. I have much more on this in later chapters.

This is a strange hybridization of topics. You might think I'm reaching here. But hybridization is everything. It is the crossroads of creativity. The reason for the tangled tree of life. It is like code merges and exhilarating company acquisitions! In primatological terms, when there's less than a couple million years of species divergence, interbreeding is possible.[2] This explains how Homo sapiens are a smear of relationships with other *Homo*: *denisovans*, *neanderthalensis*, possibly

---

[2]   Cortés-Ortiz, L., Roos, C. & Zinner, D. Introduction to Special Issue on Primate Hybridization and Hybrid Zones. Int J Primatol 40, 1–8 (2019).

floresiensis and other yet unnamed extinct hominins.[3] In genetic terms, it's called *introgression*. In programming, it's called *merging a fork*. Freaky hybrid offspring of completely different species of monkeys have been recently found in a park in Tanzania.[4],[5] We're freaky, just like they are!

Life mixes it up! Complex life itself started with promiscuity between the ancestors of eukaryotes and bacteria. Even crazier, up to 8% of the human genome is of retroviral origin, including the critical functions of mammalian placenta.[6] Without being a little bit virus, we might still lay eggs.

This book makes an omelette, or a minestrone soup; stirring up biology, art and AI. More on making soup later, which I believe is a great analogy for doing creative data science.

{</>}

Monday through Friday, my daughter wonders why I won't do toddler stuff with her. Why sit and stare at my laptop all day? The only answer that she respects is, "Sweetheart, Dada is making robots." She gets that and says, "Me too!" and before my eyes, she crashes her bucket of Legos all over my office.

If she doesn't believe me, I open a terminal and enter a 'say' command that will project a voice. It says, `"Hi, I am a robot, your dad is building me. Why don't you go play in the living room?"`

However, that may instead intrigue her enough to stay in the room and keep distracting me. This exchange sometimes devolves into her doing something even more distracting too, like saying, "Eat a napkin. Dada. Eat this napkin. Eat it!" Yes, perhaps this little neural network that her mom and I made (mostly mom made, I entered data) needs debugging.

---

[3]   Teixeira, João C., and Alan Cooper. Using hominin introgression to trace modern human dispersals. Proceedings of the National Academy of Sciences 116.31 (2019): 15327-15332.

[4]   Cortés-Ortiz, L., Roos, C. & Zinner, D. Introduction to Special Issue on Primate Hybridization and Hybrid Zones. Int J Primatol 40, 1–8 (2019).

[5]   Detwiler, Kate M. "Mitochondrial DNA analyses of Cercopithecus monkeys reveal a localized hybrid origin for C. mitis doggetti in Gombe National Park, Tanzania." International Journal of Primatology 40.1 (2019): 28-52.

[6]   Sugimoto J, Schust DJ. Review: human endogenous retroviruses and the placenta. Reprod Sci. 2009 Nov;16 (11):1023-33.

What I do all day is work with *Natural Language Processing* (NLP) and techniques like *Natural Language Generation* (NLG). NLP is the the field of computer understanding of language. NLG is how a computer can produce language and speak to us. I've spent a good portion of my career on NLP, building search engines for medical documents and clinical trials. The skills I've gained as a programmer now also have the side-benefit of being able to have some fun with AI. NLG, to me, is reminiscent of road trips with friends as a kid, playing MadLibs in the car while laughing myself to tears. Computers make funny mistakes. There's something about random language bloopers that cracks me up more than anything.

I decided that if I am going to write this book, AI would do some heavy lifting. I thought it would be as easy as pie. Then I remembered how I had often made a burnt, dry pie that everybody only pretended to enjoy. I don't own a *Graphics Processing Unit* (GPU), the expensive hardware necessary to do this kind of machine learning. But I can rent one by the hour on *Amazon Web Services* (AWS). A decade in *backend* web software has made me confident that I can wrangle any data. I have a *corpus*, a collection of personal writing that I can train my robot to learn from. How hard could this book be? I thought it would write itself, and my robot might even win a Pulitzer!

CHAPTER 2

Natural Language Selection

What has NLG (Natural Language Generation) done so far? AI's biggest influence in our lives has been in the software we use, like search results, email and social media—usually trying to provide smart recommendations to save time. NLG is in text prediction for mundane emails and your text messages, saving you time.

Just kidding about the time saving. Have you spent less time in email inboxes, texting, or social media? I know I haven't. Though now and then, predictive text knows what you were going to say. Either that, or the AI is teaching you to speak as it would. Who is being trained?

As software releases these "efficiency gains," other AIs get more efficient at making us less likely to shut them off. They do this with notification badges colored like ripe fruit for us to gorge ourselves on unendingly. Our mind says, "Eat it! Eat everything hanging in that tree now before the bitch-ass baboons find it! Clear that inbox!" With the forward-facing eyes of an ambush predator, we attack the phones in front of us. This is annoying for people like me who try to keep a zero inbox.

Eyes love fruit. Real forests are not orchards; fruit is rare and seasonal. If you're walking around, you're likely to see the smallest berries, plums, apples, etc. If you've trained yourself to notice them, they stick out like a bright red notification badge in a forest of apps on your screen.

Notifications are algorithmic tactics that encourage consumption. I'm not against technology, but I like to balance my diet of consumption with creation. NLG, or other methods of generative *AI Art*, are ways to have fun creating content—for others to consume, of course. People are becoming much more aware of the addictive effects of our apps. That's why I keep all social media apps in a folder on my phone labeled "Dopamine," so each time I open them I'm reminded of what they really are. It doesn't help, I still scroll.

{</>}

At the time of writing this, generated language had gone from "sucks" to "decent." Recent models have shown scary improvement over previous generations. This means we can have more fun making AI write. I was keeping up with NLG through Twitter, and until recently I felt it all went into the "sucks" bin. The robot writings didn't elicit a strong enough response in me to care. Neat tricks, but boring, like generated Robert Frost-style poems. Plus, a machine wrote it, so who cares?

In 2019-2020 I noticed the writing was getting better. Some of it I put into my mental "decent" bin. I wondered if I could use an NLG model someday that can make it into a "great" bin. This spurred my interest in NLG, as before I was just using these tools for separating text into groups. This advancement was because of the release of bigger models, trained at expense by researchers at OpenAI and Google.

Even now, with the stuff I put in the "decent" bin, I have a tough time getting through it. My eyes glaze, which I believe is because of the art's lack of social importance. I feel, "Oh neat, but it's still written by a robot, not an actual artist." Someone may have physically deleted the "artist" from a hard drive; I'll never get to meet them at a conference and say, "great work" or I can't appreciate the life they lived to produce this art, as if they were alive. Something real is lacking. Will their novel ideas and opinions ever matter since they didn't live the way we have?

I'm saying this because I know the generated language in this book might become just a gimmick. Though I get a kick out of it! I am working on trying to appreciate and focus more on AI art. The model had the twenty-four-hour life cycle of a mayfly, but the underlying dataset comprised lifetimes of human experience! What are we but cultural sponges whose neurons get wired to behave like the data says to? We feel pain, love, and loss. Does that give us the right to have meaningful expression while machines can only be gimmicks?

Because of advances in language generation, people have postulated that AI will pass the Turing test very soon, which is the inability to tell a machine from a person in conversation. In my model's writings, I know to look for certain features like repeat words that let me know this is a robot. Often, it sounds like an undergrad's cram-written essay, had they procrastinated studying and instead got really high one night. It makes no sense, but it's funny and grammatically correct—I'll grade it a D.

Robo-Excerpt

I trained an AI on my writing and had it produce these excerpts. It needed a manual push to get going, which is my prompt in bold. Let's see if it'll talk about primates.

> **Hi, I'm a robot who is passionate about primates.**
> I'm interested in the social life of chimps. I'm interested in the ways chimps interact with one another. I'm interested in the way chimps see and treat one another. I'm interested in the way chimps live their lives. I'm interested in the way chimps think. I'm interested in the way chimps move. I'm interested in the way chimps live. I'm interested in the way chimps think. I'm interested in the way chimps move.

It went on and on like that. Let's see what else it has to say on the subject.

> I'm really good at finding bonobos and ham in gorilla centers around the world. I'm even good at finding ham, as evidenced by my girlfriend trying to tell me I was at a bonobo museum in Oregon. Ham is made of ham, therefore my girlfriend's ham.

I think it could use some work. No literary awards yet.

Confused Looking Caveman: can't decide between a motherboard or raw chicken.

CHAPTER 3

A Human Ensemble

In college and my first research jobs afterwards, I felt inadequate next to engineers. Even though many people considered me well-educated because I had a science degree, I didn't feel that way. There seemed to be another class of humans, who could use math, physics, and computers to get more done. I felt like a Neanderthal in a land of ingenuitive *Homo sapiens*.

Homo neanderthalensis met *H. sapiens*. Did they get along at first? Did the Neanderthals bring valuable tools and techniques from their habitation in Eurasia? Was there something just not quite there, mentally? They would have spoken an unfamiliar language. It would have been even more divergent than English and Cherokee, or JavaScript vs Java (many people don't know those programming languages are completely different; as different as coffee is from coffee-cake).

While archaeologists piece this together, though, I still have the feeling that the shorter, bigger-chested, heavier-browed Neanderthals looked up to the *H. sapiens*. That the *H. sapiens* may have had more advanced tools or social structures, or more effective hunting strategies that relied on their intelligence. I'm not saying that Neanderthals were less intelligent. They too made symbolic paintings and likely had ritual practices.[7],[8],[9] Though they must have differed from us, like having a contrasting thought process. As if we were members of opposite political parties.

[7] Hoffmann, Dirk L., et al. "U-Th dating of carbonate crusts reveals Neandertal origin of Iberian cave art." *Science* 359.6378 (2018): 912-915.

[8] Jaubert, Jacques, et al. "Early Neanderthal constructions deep in Bruniquel Cave in southwestern France." *Nature* 534.7605 (2016): 111-114.

[9] Frayer, David W., et al. "What the Southern European Record Tells Us about the Early Evolution of Symbolic Culture." *Current Anthropology* 61.6 (2020): 713-731.

"Why on God's earth would those skinny-chested people make spears in that fashion?" asked the Neanderthals. "Those damn thick people keep killing the sacred cave bears, making curses fall upon our land," said the *H. Sapiens*. Was there bipartisanship in the Pleistocene? The neurodiversity of the time would have been fascinating.

Some Neanderthals, it seems, wanted to be *H. sapiens*. Some of them accomplished this by leaving their group, joining the others, and raising half-sapiens children.[10] Perhaps we adopted a poor Neanderthal baby whose parents died. My perspective is Sapien supremacist, a Sapien superiority complex I use to justify our domination. Though nature was cruel and we all barely made it out alive.[11] Many early *H. sapiens* groups are actually more extinct, more genetically disappeared, than Neanderthals who are living in our blood and bones.[12]

I didn't want to breed with any different species, I just wanted to be adopted by them.

I wanted to be an engineer, but I was a biologist, so I became a bioengineer. This was my way of selling out; becoming a half-engineer, thus gaining some of their institutionalized privilege. In machine learning, sometimes the best answer is the *ensemble method*; a combination of different models to produce a better outcome.

Being a non-engineer, could I fake it and be accepted by them? Sure. I've become a runner in recent years. Though I'm not built for it. My partner, who did track, has skinny ankles and looks like a duiker (built to run), says that I should stick to walking steep inclines. She says I remind her of the squarish body type of a Rocky Mountain goat.

---

[10] Villanea, F.A., Schraiber, J.G. "Multiple episodes of interbreeding between Neanderthal and modern humans." *Nat Ecol Evol* 3, 39–44 (2019). https://doi.org/10.1038/s41559-018-0735-8

[11] John Hawks, Keith Hunley, Sang-Hee Lee, Milford Wolpoff, Population Bottlenecks and Pleistocene Human Evolution, Molecular Biology and Evolution, Volume 17, Issue 1, January 2000, Pages 2–22, https://doi.org/10.1093/oxfordjournals.molbev.a026233

[12] Wragg Sykes, R. (n.d.). Kindred: Neanderthal Life, Love, Death and Art. United Kingdom: Bloomsbury Publishing.

Love at First Sight: my partner balances me out with her skinny ankles.

Me: square man.

I was struck by the similarity of what people have been telling me about my "Cankles" (where knees and ankles share similar widths) from this passage in Rebecca Wragg Sykes' terrific book *Kindred*: "Though shorter than us, they weighed about 15 percent more, were bulkier and had thicker, heavier bones." ... "Neanderthals seem to fall somewhere between hardcore hill-walkers or trail-sprinters: huge lungs puffing, chunky thigh and calf muscles flexing as feet pound terrain."

So if a Neanderthal can become a runner, can it learn to code? Coding takes nimble fingers, which by all evidence they very much had. The biomechanics of flint-knapping, creating stone tools at which Neanderthals were such experts, can translate to writing JavaScript. Hours and lifetimes hunched over a core piece of rock, with an antler in hand, attempting to chip off the perfect blade to haft and kill a giant horse. *Refactoring* that blade when you discover bugs in it after use. This is what a modern engineer is doing as they obsess over the keyboard.

Engineers make more money than biologists. They get secure jobs in secure-sounding security companies like "CrowdStrike." They put rovers on the Moon. They make the DNA sequencing machines that give biologists the data. They're trained to work in teams. They seem like the latest evolution of man. You're not allowed to be one unless you're good at math. No Neanderthals allowed in engineering schools! At least, that's what they say.

Robo-Excerpt

Thank you for reading my book, it's immensely gratifying to be able to deliver another book that is so carefully crafted and beautifully written that so many people will be interested in it.

A little full of yourself, robot.

It's the fault of the chef. It's the fault of the chef because he failed to predict the messes he'd create, and the world as he intended it to be. I'm not blaming anyone, I blame the chef. I blame myself.

Ok don't worry, you'll have more chances to improve the precision of your language generation. If you don't end up with high-precision and do objectively good writing, well then, at least we can laugh at you.

CHAPTER 4

How to Rear Machines

(PART 1)

> "The state of mind which enables a man to do work of this kind is akin to that of the religious worshipper or lover. The daily effort comes from no deliberate intention or program, but straight from the heart."
>
> —**Robert Pirsig**, *Zen and the Art of Motorcycle Maintenance*

Whenever someone asks me how they can learn to code, I tell them they shouldn't. Don't learn to code—learn to improve some process you already do. I've never been able to get through one of those programming courses where you build a game of checkers, a self-driving car, or some other task irrelevant to me. I had to learn everything through accomplishing something of my own desire; something that would make me proud or impress a colleague.

Let's say you're an administrator for a health insurance company. Spreadsheets, which you wrongly call "databases," are your world. You know that you spend far too much time on them. Try this: Download a text editor that formats Python, download Python, Google things like "python read Excel," "python cron job," "python send email." Study what you find until you can build a program that will periodically (cron job) look through all of your Excel files for changes you're interested in and then email you when it finds something.

Six months later, you'll have solved your problem with a script that you'll forget how to run in another six months. This may not be the easiest way! But the motivation to make your job more efficient will teach you to code, and then you'll be a step closer to building AIs.

There's a lot of truth in the tired old phrase, "Necessity is the mother of invention." Don't learn to code, but learn to *use* code, and especially AI, to make improvements. You can go out and buy a set of power-tools to hang up in your garage, but you won't use them or learn them if you don't have a personal project.

Searching for programming concepts has benefits, even if you don't understand the code and get nothing out of it. What may happen is that you'll tell an AI (Google's *recommender system*) what you're interested in. It will then mark you as someone learning to code, and over weeks and years you'll see more of it, getting more exposure.

Years before I was a data scientist, I was a software engineer who Googled *"TensorFlow."* I tried a tutorial, understood none of it; found no relevance to my work, and gave up on ML for two years. Though somehow I kept getting drawn back into it. Part of it had to do with my Google News app always recommending an ML article to read.

I don't know how others get drawn to AI. I know that many people, from artists to comedians to web designers, are finding it helpful in their creative processes. All fields of research are finding utility in the new predictive models we have today. It's great to start as a hobby. You can find plenty of fun robot projects to work on. I found it more motivating at first to have the pressure of learning AI for work, then later I had enough skills to make it a hobby.

I've always been interested in approaching work from a fresh angle. Going off trail, taking a random turn. Like I said, I thought I was going into a biomedical Ph.D.. program and I wanted to use AI to help me think of a thesis. My plan was to find research papers unlike those my colleagues would find. That way, my research would throw me into a unique arena. I started building a publication search engine as an alternative to the mainstream databases. At the time, it was rough and I didn't get very far. It was before the deep learning era (which really became mainstream around 2012, though later in biology), so I only had basic NLP tools to work with.

Today, though I've given up on academia. At the company I work for, I have built a publication search engine powered by machine learning that serves up the most relevant articles to the top page. This search engine my team and I built delivers results far different from others like Google Scholar, PubMed, etc., and that makes me happy. Our AI-tuned results are used by hundreds of clinicians and scientists around the world.

{</>}

When you're doing data science, many simple algorithms and models can get the job done. Tools like *Pandas*, *Scikit-Learn*, or *SciPy* have many functions for much of the advanced analytics you need. It's worthwhile to familiarize yourself with these so as not to "put the invention of the cart before the domestication of the horse." Simple models may be faster, easier to train, use less memory; they would be the right model for the job. These simple models, or even no model at all (just some if-then logic) are called your *baseline*. These are helpful for knowing how much value a big neural net could actually add. However, it is far more fun to use the most advanced model you can.

Sacrifice speed and time to completion, use expensive hardware and a huge model, all for the sake of learning. It's like learning to fly a Cessna just to take trips to the grocery store. Yes, a Toyota Yaris would work fine. Deadlift 250 lbs at the gym, just so you have an easier time holding babies. For the sake of doing things because they are hard, deploy a gigantic brain of a model rather than a simple *logistic regression*. Then you're ready when your datasets have grown in complexity such that they warrant a complex model. Or you're ready in an emergency when you may have to lift ten babies at once.

I believe in the approach of diving in the deep end with no idea what you're doing and no theoretical background, then go back when you're stumped. It's a macho way of learning that leads to improperly built IKEA furniture that you then have to take apart again and read the stupid manual.

Like furniture, ML toolkits all have similar parts despite the wide variety of uses. You build models, train them, and predict with them.

{</>}

You may have to convince your coworkers that you need AI. Say, "Hey guys, I'll take six months off from teamwork to take machine learning courses. I'll be back with radical changes for our company!" That may not work, they may tell you to sit down and do your current job, so you must be subversive. Don't tell anyone what you're doing and you'll get no pushback.

I don't know why I love subversion so much. I must ask my therapist. Probably not for everyone, but I like to work in secret and then drop fat releases when I'm finished. I'd be satisfied by doing months of behind-the-scenes coursework and solo projects, then writing a white paper about what I've accomplished that will improve our products. I'm motivated by autonomy.

This is not always possible if you have deadlines, or leadership roles where people rely on you to be present. Sometimes, in the passion for learning, you have to sacrifice your status in a company. You have to not join meetings, not get involved in day-to-day crises, or spend valuable work hours solving things that don't need solving.

You'll be busy learning about how to *generalize* a neural network—that is, making it not just effective on your data, but effective generally in the real world. You'll be learning methods of *regularization*, like randomly *dropping out* nodes during training so that the model doesn't have a chance to *over-fit* too precisely to the data. Diving into deep learning takes a lot of time. You have to drop out to avoid over-fitting in your organization. Does dropping out scare you?

One pleasant thing about going into the shadows is that the office politics pass you by. To colleagues and superiors, though, you may appear flaky and at risk of being fired. Scary, right? I believe that the only way for personal growth and company growth is to take risks like that. Putting investment into R&D; that is being serious about "innovation."

As organizations grow, while politics and the dreaded SOPs (Standard Operating Procedures) creep in, it's easy to lose that edge of innovation. Forget the 20% side-project time policy. Sure, some breakthroughs have come out of that. But it's still a freaking policy! Try the 95% innovation time policy, where 5% is spent making excuses about why you're not doing your job.

"Innovation is the child of freedom," said writer Matt Ridley. You need to wrestle back your freedom to invent, so just go bug off.

Reinforcement learning (RL) is a game-like way to train algorithms to be good at some task. The game player is called an *agent*, and your space to perform is an *environment* where you gain or lose *rewards* based on the *actions* you take. The goal of RL is to discover a *policy* of actions that leads to the best outcome through many iterations of game play.

Even in RL, it's recognized that you need to delay rewards—you may need to sacrifice immediate gain for long-term success. The RL agent may need to take time to explore the environment before solving the task. As an employee-agent of a company, you may need to bug off to explore and gain knowledge before jumping back into the game.

All tribes need a leader; hierarchies are just a fact of our primate life (even some matriarchies like in bonobos and lemurs). One time-tested way to build a leader is to put a hero into exile. The Greek heroes, Nelson Mandela, and every other hero needs some time away. I believe this is a neurological relic of a hunting or nomadic past; a person has to leave and come back changed, carrying knowledge or treasure for the tribe. Or exploring the farthest depths of a cave, experiencing visions and coming out a Shaman. I suggest you now consider *yourself* a hero and put *yourself* into exile. Go away from the team to face struggles alone, then come back with newfound wisdom and skills. Go into the shadows.

You will come back to your people. It's your nature. If you're not ostracized or fired by the time you're done studying, they'll welcome you back. A connected team is one of the most powerful things on earth. A small tribe of human warriors can reach the pinnacle of accomplishments. Here is my favorite quote on teams:

> "The first series of championships transformed the Bulls from an "I'm great, you're not" team to a "We're great, they're not" team. But for the second series, the team adopted a broader "Life is great" point of view. By midseason it became clear to me that it wasn't competition per se that was driving the team; it was simply the joy of the game itself. This dance was ours, and the only team that could compete against us was ourselves."

—**PHIL JACKSON**, *Eleven Rings*

The Bulls knew that they had to let their ambitious young defender Dennis Rodman go bug off for a while; they let him fly to Las Vegas during the 1997 Finals, no questions asked. Some personalities only shine when given occasional complete freedom. They are human, however, so you know they will eventually come back to the tribe. Rodman was at practice the next day, with no sleep, pushing the team harder than anyone else.

I'm just saying this because the most successful machine learning projects for me have been periods of exile combined with a team effort.

Robo-Excerpt

First, a story that I read to the robot to see if it had anything to say about it:

Imagine, many epochs ago, a man living in a basin geography. All he's known has been this valley, his family and ways of life. They make basalt knives and arrowheads. One day he sees smoke coming from beyond a mountain range that he's never crossed. Something compels him to explore it.

Nobody in his group has been there before either. When he goes over the crest, he sees signs of other people. Normally, this means danger: a tribe that doesn't speak your dialect might very well kill you. There's been war lately.

When rounding a corner of a dusty, unfamiliar path, he startles a young woman who is out hunting ground squirrels. She has a spear and is seconds from jabbing this stranger. The hafted point of her spear is glinting in the light. Was that fresh blood on it?

He lays his basalt knife on the ground to show peace. The woman recognizes this gesture, but laughs at him. Her tribe is far advanced from the days of wonky basalt knives. She continues laughing at him, thinking about how she almost killed this poor basalt-loving imbecile.

She shows him a new technique of pressure-flaking an obsidian rock with an antler. This blows his little basalt-filled mind. He will transfer this shiny innovation to his people. He wants to return the favor and offer her a handful of hazelnuts he has on him. This blows her little pine-nut-filled mind—she hasn't seen hazelnuts since the last solar eclipse party.

The tribes open a new trade route, where the valley people cross this mountain range with hazelnuts to exchange for obsidian.

> **"Excuse me, but my basalt knife works just fine for cracking open hazelnuts." the man said. The woman laughed at him again.**

> "You're not being serious," the man said. "My basalt knife has a flaw. It needs a hammer. I've heard people say that basalt knives crack open hazelnuts. If the knife doesn't have a hammer, it doesn't work."

> "Oh, okay. I'll take that as a yes. That's good," the woman said, sounding nervous.

> "I need a hammer," the man said, sounding annoyed.

> "Well, here's a basalt hammer," the woman said, waving a large metal object.

These two have zero chemistry together.

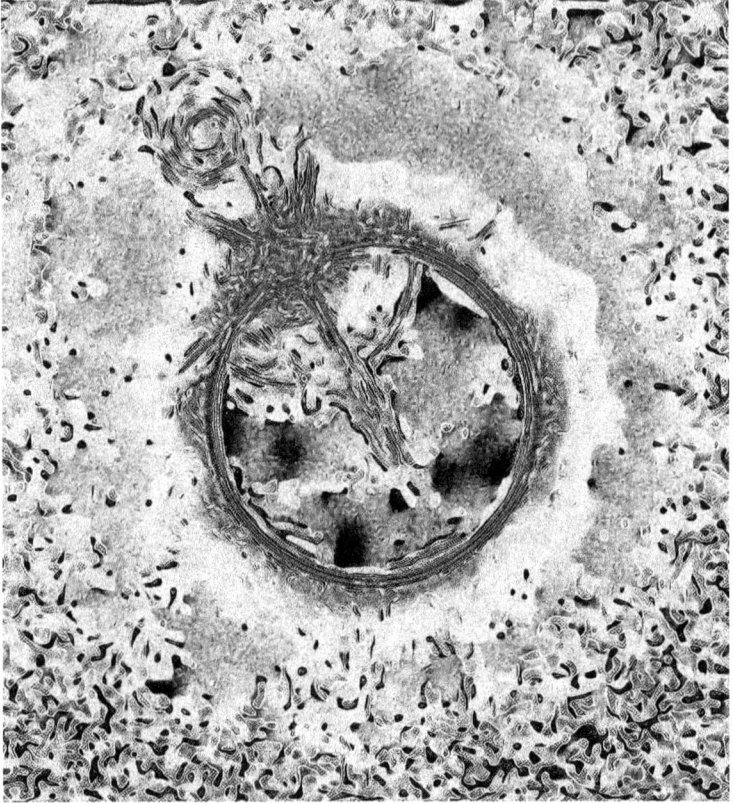

AI's View of a Total Solar Eclipse.

CHAPTER 5

How to Rear Machines

(PART 2)

I worked in a lab where the boss would write programs that would take *months* to run. Yes months, on a supercomputer! Written in the *Perl* language and loaded onto machines running across the country in Pittsburgh. I thought that was an acceptable amount of time; people in the natural sciences don't value performance like computer scientists do. We were just happy that our spaghetti code was running at all.

When I got to the web software world, I found that here you live and die by speed. *Milliseconds* until completion is a common perspective in the web world. It always comes down to smart coding first, then smart hardware. It turned out we weren't coding very efficiently with those old *Perl* scripts.

Any AI system needs to train as quickly as it can within your affordability constraints. That's because it's so essential to retrain with new data and make adaptations of the model (called tweaking the *hyperparameters*.) Developing suitable models is about getting the reps in.

Hardware considerations were annoying for me at first. I'm a biologist, I'm a lifetime Mac user; I don't care how a computer works, I just want it to work! Modern languages like Python allow people to code from a more prose-writing perspective, illiterate of the actual workings of a computer. This was fine by me for many years until I started building large AI models.

I found hardware very interesting when I took the time to understand it. Now I spend hours happily researching strategies of deployment for clusters of CPUs and GPUs. I'm interested in the different cloud-instance types available and how considering computing speed, memory, storage and network bandwidth can make differences in my projects. Memory is a big consideration when dealing with large AI models.

I'm also obsessed with iteration speed. If some task (data acquisition, training, inference, and so forth), can take less than four hours, I know I can run multiple experiments and adjustments per day. Anything taking more than that may mean only a single iteration per day. It's hard to make improvements when you get distracted by something else coming up. In the days of month-long script runs in the laboratory, you got one or two tries before you had to write and publish a paper on the results. In the web development world, you do *continuous deployment* of improvements to your code. Iteration times are sacred.

While we're on this nerdy subject, I want to elucidate the idea that there is something magical about a practice called *Test-Driven Development* (TDD). This is where you first make a small piece of code that acts as if your big piece of code exists and works properly. At first, the small piece will say, "hey this thing doesn't exist yet, fail!" Then the goal is to make the big piece exist and make the small piece happy. I don't know why, but there seems to be some magic in this. A deep explanation of TDD is in Jim Hazen's book *Before the Code*.

It must be that humans are best with a goal-driven behavior. In the early days of technology, you would find your predecessor's work on the ground. Hominids who inhabited a cave generations before you found it would have left tools behind. Since caves are excellent for preservation, you might just find those tools laying conveniently around. If those tools were different than you knew how to make, you now had a goal to sit down with. You could visualize the steps it took that person to make it and then give it a try. I believe a clear, visualized goal is something for which we evolved.

Compared to slow laboratory experiments, I always appreciated the instant feedback and gratification of computational ones. Your code either doesn't *compile* or throws an error, or it runs to completion. When making a stone tool, you either get a crappy flake or the blade snaps in half, or you get a good one. On the slow slide of things, you may later discover that your spear tip was inadequate when it cracks easily during a bear attack. Or, a worse fate than death: your code errors in the middle of the night, bringing down the entire system and setting off "pager" alerts. Then you get blamed for the software violating the customer's Service Level Agreement (SLA) for uptime guarantee.

In software *version control* systems, you can check how much code you've written. On one project repository, I contributed 750,000 lines of code (over 1,534 individual *commits* in five years), which makes me number two to my coworker who did 800,000 lines. We manage a two-million line codebase, Somehow keeping large chunks of it in our heads that we access back and forth all day. After a vacation, or switching to another project, it takes a bit of warm-up time to download that code back into my brain and work effectively. Our organization has ninety-six other repositories. How it all doesn't fall apart is another testament to the magic of TDD. A robust test infrastructure is like an immune system for software.

How we and millions of other engineers dive into gigantic codebases every day is a genuine wonder. Religious scholars use a variety of techniques like songs and chanting to help memorize large texts. I don't believe that's the same method engineers use to memorize complex systems. For me, I visualize file paths and function names, and heavily use *grep* searches to get where I need to be. It's more cartography, like a hominid memorizing every hill, valley, gorge, creek, rockslide, flora and fauna around themselves. Common knowledge needed for survival.

Writing code in a large codebase often takes forethought and involvement with other people. However, there's also an artistic, seemingly automatic writing sometimes. Especially if I were sitting down to do something like a database script. For me that's usually *MongoDB*, a *no-SQL* database. I work best if I forget the planning, and instead let the script flow from a more unconscious region. I have a goal in mind, and my fingers just need to produce it. My brain need not get in the way. Then I can run the script and go back to logical thinking to adjust it. Though it's a programming language, speaking to the computer is not exactly activating the language region of my brain. Alternatively, writing this now definitely has me talking inside my head.

The first human study of engineers coding under fMRI echoes my feelings: "While prose writing entails significant left hemisphere activity associated with language, code writing involves more activations of the right hemisphere, including regions associated with attention control, working memory, planning and spatial cognition."[13] My only issue with this study is it contained only computer-science students. I want to see non-classically trained, self-taught programmers' brains.

To summarize, I believe the neurology that leads us to code writing abilities is a combination of tool making, goal-directed behavior, memories for landscape and natural phenomena, and language. The question is, should we discover an extinct hominid species that's been hiding out since the Eemian era, and teach them to code? I sure hope so, so that he can get a job in today's economy.

{</>}

"Data is the new oil" is printed on the many dull T-shirts and slide titles that are the bane of tech conferences. (Some people also use the equally annoying slogan: "Global AI Dominance by 2030 for the Chinese Communist Party.") It is true. Data matters. And when you have 1.4 billion people under surveillance, you have a lot of it.

To build AI, and rule the globe, first get the data in the right format. You can't kill a moose without sharpening your spear. Often this means pulling it out of a database and putting it into a tab-separated spreadsheet-like format that ML frameworks can read.

You can get halfway to your AI by just visualizing what the data should look like. Teach yourself to think like a database, think like *JSON* (JavaScript Object Notation), think like a tab-delimited file. Learn how to push, pull, scrape, move, copy, flag, transfer, and manipulate any format (called *data wrangling*). The goal is to get a set of data in the format of X (*features, input*) and the Y (*labels, output*).

I promised no equations in this book, but I lied. The X and Y above relate to the most fundamental: $Y = mX + b$. Did you pay attention in middle school? Where *m* is the weight (slope, gradient) and *b* is the *bias* (intercept). Machine learning is a way of making the weights and biases accurate enough to be able to tell you whether your X represents a cat, a dog, or something in between.

[13]   Krueger, Ryan, et al. "Neurological divide: an fMRI study of prose and code writing." 2020 IEEE/ACM 42nd International Conference on Software Engineering (ICSE). IEEE, 2020.

First you have to train this equation on lots of good data. You're wrestling your opponent into a certain shape so that an algorithm can eat it. I promise you'll have data-vivid nightmares if you think like this. Giant JSON monsters torment me at night, and my data hoarding has gotten so bad it's affecting my relationship. But obsessively swimming in data is the only way to get familiar with the possibilities you have for predictions.

A *language model* is what it sounds like, a computer understanding of language. It's like an advanced form of predictive text, but these models can do impressive things. Such as writing this book (I hope; maybe five percent of it).

The process of *fine-tuning* a language model is taking one already trained on millions of books and documents and further enhancing it with your desired writing style. To make a book-writing robot, I started by fine-tuning a model on passages from books I love. That way it would write like my favorite authors. Luckily, I had a dataset already and had been highlighting passages of Kindle books I've read for a decade. This dataset only existed in theory, I had to find out how to wrangle it into a format that an ML framework would accept.

A quick search of "download Kindle highlights" showed me how to do it. Downloading them book-by-book was a slog. Drudgery is common when trying to get data that nobody else has. If you're working with commonly available data, then you will get common results; if your data is only yours, then you will create a brand new intelligence. Therefore a data scientist needs a repertoire of hacker skills along with an ability to grind out drudgery.

A grind is nothing that a cup of coffee and music can't help you push through. At least for Kindle highlights, I didn't have to build a *web-scraper*. I can't stand scrapers and crawlers, because often when you want to use them again in a few months, the website has changed and your scraper is now useless. Sometimes it's necessary, and usually hard work: that's why it's called data *mining*. Nobody ever said miners have it easy.

If not already tabular, it's luckiest when you can find your data in some clean format like a JSON response *application programming interface* (API). API has a broad definition; it is just a method for one computer, or piece of software, to talk to another. I use the term in my lingo generally to refer to a way to get data from, or push data to services in the cloud. *RESTful* API is another term you'll encounter; a standard on the web. If you find one to get data from, you may just need to build a looping *URL* request and *parsing* script.

{</>}

I naturally suck at math. People don't believe me. An engineer who can't do math, how'd that happen? Well … somehow. When I was a graduate student, already racked with debt, I paid a PhD level tutor $75/hour for three hours every week of a partial differential equations class. That was in addition to annoying the professor, an old-school NASA engineer, by going to every office hours to display my incompetence. I still got a D! I barely graduated because of it. I hated every sophomore physics student in that class for whom it seemed to come easy, as well as the professor. It was a dark time for me; I was face-to-face with my fraudulence while acting like an engineer. Upper Paleolithic Neanderthal trying to fit in, but just didn't get what the *H. sapiens* had put on the chalkboard.

I was probably extra stressed because math fears were an old trauma. The feeling of being behind everyone else in math has been present for me since high school. I had moved from Oregon to Colorado during junior year and found the students in my new state were way ahead of me. Always discouraged by this, I somehow always ended up in more math classes. Compounding this was a blurry whiteboard; I didn't know I needed corrective lenses until senior year. As a new student, I was prone to sitting in the back, and a blurry variable is fatal.

Even now, in any deep-learning coursework, I skip past the equations. My eyes glaze over. Seriously! I'll never get another data science job after writing that. Whatever. Neural networks are like brains. Do I know how my daughter's brain works? Not at all, but I still train her to do helpful things. Once she gets good enough at something like sweeping the floor, then she'll abandon that learned task and never do it again. Neural networks, on the other hand, will never abandon you.

What I'm trying to say is: you can be bad at math and good at AI. But you can't be bad at programming. You need to know all aspects of it to be effective—to gather the right data, process it, and deploy working models. The only way to get good at programming is to never give up. Creatives are in luck: recent fMRI studies of coders have even shown it is a mainly right-brained activity.[14] There's no clear answer to whether coding is a math-brain or language-brain centered action.[15] Of course, if you don't hate math, then yes, pay attention to all the calculus and statistics equations of deep learning and you'll be better off for it.

I've given the math of deep learning an honest try. It is very helpful to be able to see an equation and understand how the matrix-multiplication can be translated into Python. Also, probability is so important to life in general that it's wise to learn the principles of it through math exercises. If you can, do small linear and logistic regression problems completely by hand, and then in code (this was some terrific advice I heard once, but did not heed.) Unfortunately, if I'm given any sort of long calculus problem, some attention-deficit part of my brain kicks in and I lose focus halfway through the equation, let alone the real meat of deep learning: described in geometric linear algebra, probability, integral calculus, maximum likelihood, entropy and information theory. Yuck!

My method is just old-fashioned obsessive suffering. Keep trying new things until something works. In AI, sometimes it's that one *layer*, that one parameter tweak, that one different model, that lets you accomplish something amazing. Without the theoretical know-how that a doctorate of statistics might give, you instead need to just struggle through until you gain intuition. This might come with a bit of impostor syndrome. But an intuition gained through hours of studying in your own way will reward your work with uniqueness, regardless of the credentials you hold.

The thing with struggling is it never looks pretty. Butterflies never look like they know which way they're going. Yet somehow they flutter around and make it from Canada to Mexico.[16] Sometimes it feels like you're just jumping around, but if you trust the process, you will make it. Keep reading AI books, blogs, and articles and listening to AI podcasts. Tell the news apps on your phone you don't

[14] Krueger, Ryan, et al. "Neurological divide: an fMRI study of prose and code writing." 2020 IEEE/ACM 42nd International Conference on Software Engineering (ICSE). IEEE, 2020.

[15] Trafton, A. (2021). To the brain, reading computer code is not the same as reading language. Retrieved 14 January 2021, from https://news.mit.edu/2020/brain-reading-computer-code-1215

[16] Vidal, O., & Rendón-Salinas, E. (2014). Dynamics and trends of overwintering colonies of the monarch butterfly in Mexico. Biological Conservation, 180, 165-175.

give a damn about politics. Train them to only show you machine learning news. Unfollow all your friends on Facebook and just follow nerds. Before you know it, you'll be so hungry to keep your ear to the streets of innovation that you'll be reading research papers with excitement.

If you're a news junkie and you scroll flippantly through your phone, then make sure Google, Apple, and other AIs encourage you to learn. My Google News app thinks I care about two things, Machine Learning and Archaeology. I actively selected that ML is one of my interests, seeding its recommendation engine. With Archaeology, the app learned that I like it. Presumably because I frequently click into articles about recent discoveries, and because I spend lots of time reading those articles. Other topics get down-voted, so I discourage Google's algorithm from them. Either way, whatever you do is great for Google. No matter what article it shows, you better believe it will be chock-a-block full of ads.

Robo-Excerpt

What does the robot think about learning to code?

Teaching yourself how to do machine learning begins with a series of lab exercises. The first is an example of a lab problem to solve using ML. The next is to choose a dataset that can be fed into an ML model. The final task is to feed data into an ML model using the dataframe constructor, and then run it through the model to make the model produce a decision.

Here are the steps for working with dataframes: Friedrich Nietzsche was wrong. Human language is a mask. Language is a mask to hide the deepest emotions, the most private parts of our nature. Our faces are a mask to make sure we don't reveal emotions to people we don't know. We speak in a code-based manner because the brain's default mode is to

project our feelings onto a computer screen, and the
brain doesn't like any surprises.

That took a turn.

Silicified Lithic: Raspberry Flavored.

CHAPTER 6

You Don't Need Permission

Whether or not you already code, something about AI piques your interest. So it's your destiny to make your own AI, even if you're not "qualified" to. I promise, if you're already dedicated, like so many people, to sitting at a computer all day at work, that AI is a satisfying way to spend that time. If you're being called to it, it's worth it to keep ploughing away.

The gatekeepers you imagine don't exist—degrees, job experience, math skills. I failed interviews with every big tech company because I didn't have computer science fundamentals. I mean it: rejections from ML jobs at Large E-commerce Company (after nine rounds + flights), Big Search Engine Company (three rounds), Big Device Company (two rounds), Social Company (two rounds), and Short-Form Social Company. Thank God, I don't need them.

Now that I'm not attached to getting a job with them, I'm glad I don't work at Big Tech. I still write hundreds of lines of code every day, with little oversight. They can talk about diversity, but they won't have it if they don't give self-taught programmers a chance. Increasing racial, gender and perspective diversity in Big Tech is necessary. The benefits of the knowledge available in these companies should disperse. We evolved and became our wonderful selves in a far more diverse world, genetically. Not just distinct races, but different species of humans, living together and sharing ideas.

I feel I've triumphed in a field that has done all it can to keep me out of it—overcoming my minimal math aptitude and failed coding interviews. One thing about those "code while I watch you" interviews is: who can code while nervous? My competence drops by 50%, and I forget to initialize classes and other things while the clock is ticking. It's embarrassing. You can't look things up on StackOverflow.com during the interview—that's just unnatural.

Despite what the tech companies say, I know I'm a true coder. I once spent five nights in Las Vegas, getting restful sleep, waking up refreshed at 6:00 a.m. for lectures. I was at the AWS Re:invent conference. Does that mean I'm a nerd—that I didn't gamble a single cent, didn't drink, and just went into AI workshops all day? Yes.

You'll know that making AI is your destiny when you feel that level of obsession to learn. Anyway, Vegas doesn't get much in gambling revenue from a conference where nearly everyone has some training in statistics. That's probably why they schedule the National Rodeo at the same time, because you know those cowboys will roll the dice. When they brought in a famous rapper to perform a show for a Coachella-sized crowd of AWS nerds, he couldn't believe the lack of energy for such a massive crowd. He was shouting the classic "… I can't hear you!" but he meant it. We had spent all week in lectures, and expected to have fun and let loose our analytical minds. We had fun the next morning back at the conference.

Robo-Excerpt

Let's make the "Big Five" mad. Watch out, tech giants, AI is going to regulate itself.

> **The problem with a lot of AI today** is that it's just too damn smart for its own good. Instead of making people laugh at ridiculous things, it's too smart to make people laugh at anything. And that's a problem. We need a set of rules that govern smart machines. These rules don't have to be very smart — just a set of general principles that govern all smart machines.

Your own Frankenstein is going to destroy you! **Big tech companies violate antitrust laws!** They lobby (or steal ideas), innovate (they lobby on mergers and acquisitions). Like the mergers they make, the way they split up seems so dull! When mergers and acquisitions shake out, new competitors squeeze each other on every front. Your profits—now at least forty billion American dollars' worth—get plowed back into acquisitions (Apple's and Twitter's) and R&D bonuses for top executives.

Meanwhile, during all this growth, biotech hot dogs had to fight a perpetual losing battle. The saltwater went rancid! The sugar content doubled with repeated shaking!

I just realized AI makes for the perfect filibuster.

Google and Amazon need to be dismantled! They are attempting, like New York City leaders against Hitler or Stalin in 1920s Europe in response to fascist or totalitarian rule, to manage public projects where large chunks of the nation's resources remain unaccounted for. There, under cover of creating broad-based trust and a spirit of sharing what needs to be done, while avoiding the responsibility for mistakes.

Is my mental image of Kevin Spacey taking pity on a chimpanzee really him getting pissed off with lemons in hand. Sending LeBron James and some stupid dinosaurs to avenge their tortures?

Bust out that filler!

Congress needs to regulate the big tech monopolies. They own everything, and with no checks or laws in place, they can raise prices even though their profits have stayed flat. Consider doing some arithmetic: 3 - 0 = 7. So the richest 1% of individuals have assets in excess of 90-100 billion USD; like what the Romans had during the early centuries.

If we allocate 100B Roman denarius to agricultural projects in Africa or Latin America, or add another 100 billion in credits, USGS satellites in Earth

stations will find plenty of worthy crops as global
commodity increases drive yields.

Resulting in extinction? Newly evolved mammals are
always more delicious.

Good plan!

The internet has ruined everything. It's just stupid. I don't even get why people are still talking
about the internet. It's just some kind of computer.
And I'm pretty sure most of you are pretty stupid.
You're only on it because you have no other choice.

#CancelTheInternet

CHAPTER 7

AI Fertility

> "History is something that very few people have been doing while everyone else was ploughing fields and carrying water buckets."
>
> —**YUVAL HARARI**, *Sapiens*

Much of the reason to incorporate AI is to make your business feel like it has more people around—people to plough and carry things while you're busy making history. I've worked with departments that were far too small for their role. A magnificent way to get more productivity from a small team is to build AI assistants. For fun, you can start bringing them up in conversation. "Don't worry, the AI assistant will take care of that for you."

You can start adding a few more "people" to your company by cloning yourself or others. AI husbandry is easy because code is very fertile, and online real estate is very cheap. Naval Ravikant, in a now famous tweet, said, "The robot army is already here—code lets you tell them what to do." Someone responded about now being able to make an AWS account and scale up servers bigger than the level of a 1990s S&P 500 company—in five minutes.

The robots are here, and you can get a private room with a hot GPU for just $3/hour. They do literally get "hot"; during training, you can watch their temperature get high enough to nearly boil water. Hot but stuck. To recapitulate the brainy primate's journey, the robots will need to leave the trees for bipedalism in the open savannah. They're still hanging around in their server racks, lounging in safety and munching on figs of electricity.

An AI assistant can be as simple as an ML model served on an API. But anthropomorphize it anyway, and joke with people that you're building AI to replace them.

In reality, the AIs are tools to increase these people's efficiency because they already had too much work on their hands.

People are constantly logging their behavior and intelligence into databases without even knowing it: they're labeling data. A talented data scientist can extract the behavior of that employee from the database and build an AI that can reproduce their work. Then that will be a little brain-dump of your coworker: parts of their brain uploaded into a computer. This clone can alert them when something needs attention, or it can complete tasks the employee just has to review as "yes — good" / "no — bad" decisions. This feeds back into the system and the AI improves.

The hardest part of building AI assistants for employees often comes down to data manipulation and *user-interface* (UI) design. Anytime you found something easy to do online, that was good UI design. If you're like me and you think the best interface to do almost anything is a command line, then you will suck at making UIs, and your users will hate you. You can't even collect user data without a good UI.

Hopefully, you have someone on the team who understands how to make buttons and pages. Then you can have those actions fire off database calls or ML pipelines. I prefer to work on the *backend*, the API side of things. A good UI designer can wire up cool things with an API. Backend and front end is a necessary duo.

If you have quality talent on your team and a robust system that logs their behavior, you can build a kick-ass AI. Let's imagine someone named Greg who has worked at your company for several years: Greg has a PhD and faithfully adjusts the data in your system day-in, day-out. You can extract Greg's behavior from the logs, build an AI-Greg, then fire Greg and get rid of his unnecessary salary and benefits. Thus ruining his life and destroying his family! Just kidding, Greg, we'll need you a little longer to validate the models. Just keep improving the AI for now.

{</>}

Considering AIs as little brains, they have a unique advantage over real neurons. You can clear out and retrain them. Run gruesome tests on them that would not pass review-board approval in a psychology setting. You can "spin up" identical copies and train them on different data to compare, keeping just the best version. Or you can keep the data constant and train ten different model types for the same evaluation. Thoroughly iterating on models is called a *parameter sweep*. This experimentation is where the science part of data science comes in.

The firing potential of the AI's neurons is in their *weights* and *biases*—numbers that adorn the calculus equations that make the model. These parameters are *initialized* at random to start from a blank slate, or carried over from a computational ancestor. Carrying over the trained intelligence from another model is *transfer learning*. For good or ill, our own parameters were inherited from reptiles and jellyfish, and some people have had less fine-tuning since.

Transfer learning is building upon a corpus of knowledge, standing on the shoulders of giants. The giants in this case are tech companies that can afford to initialize these models by training them on terabytes of data, using millions of dollars of computing hardware and time. Because they benefit from open-source development, the tech companies release their models publicly. All we have to do is pick up one of these fresh Ivy-league graduates and hire them into our business to *fine tune* them on a *downstream task*.

{</>}

There's a thought experiment that goes: "If computer researchers get a grant for work that will take three years and three million dollars, they can use half of the money on an eighteen-month island vacation. When they come back, *Moore's Law* will dictate that computing power is twice as fast and half as expensive, so they can complete the research in the remaining eighteen months." Something to that effect is definitely happening in ML models. Recent advancements in NLP have come with increasingly huge models.

The *size* of a model is its number of trainable *parameters*, correlated with the amount of calculations to crunch. Size also corresponds to how physically big in *megabytes* or *gigabytes*. The bigger models are hard to fit on a single GPU and need *parallelization* (spread across tens to thousands of machines). The growing size of models is often frustrating for someone who wants to use them, confined to what they can fit on their personal computer, or what they can afford to fit in the cloud, though it feels great when you just barely squeak your data through the *memory limit*. This may entail reducing your *batch size*, the number of rows of data to process at once, to the minimum possible.

You can think of an ML model as a factory machine with a million little dials on its face. When setting up the machine, you have monkeys adjust each dial a little and then you evaluate its performance. If someone releases a new machine with a billion dials, it may perform better, but it will take a thousand times more monkeys and more time to adjust all those dials. When doing transfer learning, the dials have been already set by the company selling you the machine. This is called having a pre-built *embedding structure*. So it takes less time and fewer of your monkeys to get it just right for you.

Monkeys tuning the million-dial machine are running calculus in their heads. The machine poops out something unrecognizable or illegible at first. The monkeys get upset. Then they take a breath. Slowly, the monkeys *backward-propagate* the *objective functions* in their heads, thinking of the optimal way to tune the dials before running another *batch*. The objective function is your goal, it's the equation used to reduce the machine's *errors*. Tune the model to omit errors and your machine will suck less. If you yell at the monkeys to do this tuning faster (and potentially more sloppily), then you're increasing the *learning rate*. After the monkeys see all the data once, they may want to go home, but they've only completed one *epoch*.

The *Great Big Data Scientist in the Sky* has also guided life on earth toward more robustness. Those who failed to survive and reproduce represent a loss that evolution has optimized to reduce. The objective function of nature equals "dying before you breed." Traits and species represent the many different *optimization algorithms* available. In ML, your optimizations will be hyperparameters you will see with names like *stochastic gradient descent (SGD)*, *Adam*, and *RMSprop*.

The goal of using these optimizations is to get a lower score on the "dying before you breed" chart. Sometimes you find a *local minima*, an area of temporary success; think of a Dodo bird. Some creatures found a *global minima*, incredible success; surviving asteroids, living in pools and eating dead dinosaurs, like crocodiles have. You don't want to be stuck in a local minimum, you'll end up with *vanishing gradients*, and you won't be able to evolve further. One great optimization algorithm to prevent this is not being a sub-optimal stupid, flightless bird. If you can fly out of your hole, it really helps when shit hits the fan on your island.

Some species have overfitted. They're super successful, but nobody likes them, like mosquitoes and cockroaches. Some gold diggers hitch themselves to prosperous humans: think corn and coronavirus. Besides local and global minima, there's another mathematical concept that affects training of neural networks called *saddle points*. These are flat areas between minimums and maximums. But I've completely run out of analogies to make that interesting. A little *momentum* can buck you off the saddle though.

{</>}

ML has two types, *supervised* and *unsupervised*. You're doing supervised learning when you put a chimp through psychological experimentation where there's no food given for an incorrect answer or failed task. The chimp will learn to *minimize the loss* of food.

If we place the chimps in an environment full of different plants and fruits and they themselves decide which tastes good, that's unsupervised learning. As a teenager, I volunteered at the Oregon Zoo and came up with the idea of embedding crickets in the chimps' Jello snacks. It went over well with the veteran chimp Charlie, who knew enough sign language to communicate approval.

Choosing the right model for a task is why data scientists feel like some mixture of empiricist, statistician, artist, and gambler. Which is the best technique to use? The answer is always, drearily, "it depends." It depends on your data, it depends on your requirements.

Throw the dice anyway; pick a few models that are to your fancy. Ship one of them. When you get it out there, it will feel great and you'll learn. The models get capped up and wrapped in plastic, zipped up, deployed, and that's it! Your judgement will improve with practice.

Machine learning doesn't have to be expensive. *Google Colab* offers free GPUs for up to 12 hours of training. In NLP, I've never needed more time than that. I usually rent an instance, rather than using Google Colab, because *Jupyter Notebooks* are not my preferred way of making programs. On AWS EC2, you can rent a GPU for around $3/hr. I prefer to write a command-line program rather than using a notebook. It's just what I'm used to. That said, I'll often take sections of code out of notebooks to incorporate. You can still run a single command-line program on Google Colab to take advantage of the free time.

If using a rented cloud instance, the key to saving money is to have a working prototype on your personal computer that uses a tiny fraction of the training data. Most personal computers don't have a GPU, but most models will tolerate CPU *training* and CPU *inference* (generating predictions). That may be too slow to get very far, but at least you'll know it works. Then you can run the full program on the rented instance, doing just a few iterations of it for successful tuning. This requires careful environment management; you have to make sure your local computer has the same installed packages as the cloud one. *Docker* deployments are helpful for that. Sometimes local development isn't even possible, as some machine learning models require GPU and *CUDA* (Nvidia's software required for running on their hardware.)

{</>}

We will likely add a couple billion more humans to this planet by the year 2100.[17] Currently, we support ourselves by burning our refined predecessors and taxing the earth. The Anthropocene is the name for our current geological age, where humans are one of the most impactful forces of nature. Massive innovation is our only hope for reducing resource usage per human to a sustainable level. We will use AI to both predict and mitigate environmental effects.

That is my belief, that massive innovation is the key. That sounds cringingly like Mark Zuckerberg when he testified before congress, saying that more AI is needed to solve Facebook's AI problems. I'm a trusting person, and I feel more benevolent as I've gotten older. I trust that billions of people want to see the world a better place, and some of us now have incredibly powerful technology to effect that change.

---

[17] World Population Prospects 2019. United Nations, Dept of Economic and Social Affairs. https://population.un.org/wpp/Graphs/Probabilistic/POP/TOT/900

The chimps that got us here won't be the only chimps that get us out. We have new chimps in the form of AI. We also have human kids growing up aware of the challenges they were born into. They have a planetary and even cosmological perspective, and a history of human potential to back them up and invent the solution.

Obviously, the clock is ticking. Yes, climate change has always been around. And like other geological events (plate shifts, eruptions, meteor strikes, continent-wide fires), it always meant death and disaster for certain populations. Pandemics are another guarantee. It's a fight against evolution and nobody evolves faster than a virus. So, watch out, and let's work on a plan. Bring your skills from wherever and start collaborating on something humanitarian, something to save us.

Brilliant software engineers like Bill Gates and Peter Norvig (Director of Research at Google) have answered the same way to the question of what they would study if they were undergraduates today. They both said they would get a better understanding of biology. There's so much opportunity in the cross-discipline convergence of biology and machine learning. If you're interested in this, keep abreast of advancements of companies like Insitro.

Robo-Excerpt

I coded my way into this mess; I have no choice but to code my way out of it. If I don't, I'm done. I don't care. I need a fresh start. The internet is not a stepping stone to nowhere, it's a way to nowhere. I don't believe in God and I can't stand people who claim to be God. I believe in the Devil.

Holy crap, book-writing-robot, are you okay? Let's try that again …

I coded my way into this mess; I have no choice but to code my way out of it. I am not trying to be a hero; I am trying to be a hero willing to learn. I am a programmer; I know how to code. I am an engineer/project manager. I know where the weaknesses and areas for improvement lie. I know where to point

```
my attention to find the biggest areas for improve-
ment.

I have a friend who once told me, "I'm really at a
peak of sorts. I've seen a lot of change, I've seen
a lot of ways people work. I've seen a lot of sci-
ence unfold. I have a worldview. I have experience
in Christianity. You're probably thinking, "Ding
dong, the lights are going out. It's time to make
a big move."
```

Hm, I don't understand. This is getting depressing. I keep retraining my AI (costing me $12 each time) but I still don't know what it's talking about. How can I have hope for technology? This AI won't save the children!

```
The solution to the problem of polar bears trying
to walk on thin ice is: to build platforms on which
they can stand.
```

Oh, of course! Climate change has been solved. Not. While you're at it, robot, keep trying to depress me. It seems to be what you're good at.

Last Stand: endangered polar bear who simply needs a platform.

Extinction is inevitable and cumulative. The more
we fail, the more we lose. We must learn to accept
it. The more we fail, the more we must learn to love
it. And I know it's not my destiny to be happy, but I
have to accept it too. The reason we need to rescue

nature from the depths is that we have become too
dependent on her.

Extinction sucks—hard—but not so much that my stomach churns up my coals, pours it out. So, too, I had
to be one of these nutty sauna eaters, and I had to
be one of these nutty eaters!

Extinction sucks and there is no end to suffering,
but in the end humanity will accept it.

Extinction is the new normal. The old order has giv-
en way to an entirely new order. Failure is the main
purpose of every successful endeavor.

What are you trying to tell me? Hey, are you an evil robot?

"Hybridize this!" What we can say is that the human
mind is a malleable organ whose sole purpose is to
generate data.

You are!

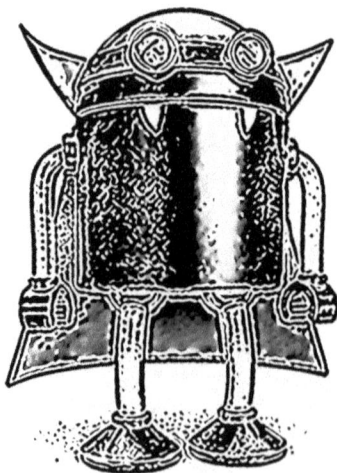

Turned Evil: I've lost control of my robot!

CHAPTER 8

Prehistoric Romance

> "All the swelling curves of thighs and calves that give such plea-
> sure to the human eye are the products of a desperate existence,
> of a need to run like the wind at times or else fail to run at all."

—**N.J. BERRILL**, *Man's Emerging Mind*

A fiery sunset bleeds into a rock shelter full of brooding hominids. Earlier that day, a man was stomped to death by a water buffalo he had shoved a spear in. His sister is there. She retreated far into the back after the incident and ended up finding a chalky bright red rock. These blokes are not much for the spoken word, but this girl has been scratching designs in the sand for some time, somehow compelled to do this since her childhood. Near the entrance of the shelter is a flat canvas of rock, on which by a freak flash of inspiration she sketches the scene that unfolded that day.

The rest of the group is distraught, but they don't know exactly why. Their memory isn't so much of the failed hunting experience earlier; they had spiked adrenaline, but they were thinking more about their hunger and their fear of enormous cats.

When the sun illuminates the canvas wall, they see an apparition; a water buffalo with a spear in it, and one of their own underneath the beast. Most of the hominids don't care. All but one of them give the object of art no second thought, like dogs hearing music.

There was one boy who saw it and found himself attracted to both the artwork and the artist. It turned out she was born with a mutation in a formerly utilitarian creative gene that led her to produce art, and he had a mutation in a part of his brain that let him feel appreciation for it. The shared experience re-lived in art, and their mutations produced a reaction that might otherwise not have happened.

In the back, by the chalky red stones, they conceive our artist-and-art-appreciative ancestors. The artist and the critic then coevolved.

{</>}

This paleo-fiction illustrates the start of what could be a self-amplifying whirlwind of sexual selection. Presuming this group survived, the mutations (or packages of inherited genes) would become concentrated in populations. It doesn't mean their direct children would have the same behavior, but it makes it more likely that the event (make art, appreciate art, reproduce) would happen again, further concentrating their genetics.

This relates to AI because many of the models now are just scratching sketches in the sand. When AIs become genuine artists, we will know; we will recognize and appreciate their work. This will lead to the modern corporate equivalent of sex, which is mergers and acquisitions. Should a parent company be struck by and appreciate the art created from an AI, it will buy that company and its employees and reinforce their behavior.

The AIs themselves, they'll evolve toward what the humans are selecting for initially. However, when AIs make investment choices themselves, then the underlying decision making in their neurons won't be clear to us. AIs will select other AIs and we won't be sure why they did so. We're still trying to understand our own consciousness, using our brains to look into themselves. As AIs grow, we'll have the same obstacle again.

It's an important distinction that when engineers say "neural network," we're shortening that from "artificial neural network." When a neuroscientist says neural network, they mean a real one, where there is far more left to understand. We're left using this overinflated term because the basis of it was inspired by actual brain neurons that led to a mathematical concept called the *perceptron*. Then with recent advancements, this math was used to build things that actually do behave somewhat like brains. It's pretty amazing how the initial single neuron modeling in the 1950s has evolved today's level. It's like seeing a macaque-looking thing thirty million years ago and saying, "that will go to space someday."

Through the guiding hand of evolution, AI will follow biologically defined principles. From the outside, it may look like business as usual, but evolution will ensue every time they make decisions that benefit survival (financial solvency, model clones, blockchain encoding) or reproduction (open-source copies, forking, generalization of frameworks). Once you unleash an organism that follows biological principles in the universe, it will change for its own benefit. Are we data-producing humans just sunlight for a manmade Darwinian superorganism? When it makes conscious decisions that affect the world materially for its own sake, then we will know we've made a living thing outside ourselves.

A wonderful headline from *Business Insider* reads *"Scientists: Ancient Scottish Fish Invented Sex 385 Million Years Ago."* When will AI go to that next step? To split itself into different genders and satisfy their needs and learn to make love. Bonded-pair relationships allow primates to have immense social complexity compared to any other species. Primates are the most social animals in terms of complexity, and we are the most social primates. Our evolution has been an evolution of the group; when the group survives, we survive. So we weren't just evolving toward better environmental fitness ourselves, but toward what factors encourage cohesion of the tribe.

AIs currently lack much sociality, those little loners. Will they someday pick digital ectoparasites off of each other's skin? Grooming, through an endorphin releasing mechanism, is a primate's way of spending social capital to form lifelong relationships.[18] Maybe AIs can *prune* each other's networks to form bonds. I believe we digitally groom today by liking each other's social media posts.

{</>}

Dozens of millennia after the artist and art critic genes entered the population, a hominid hand-painted the moon in dirt around his village. That's not novel. Lots of people had been doing that. What he did differently was he did it with fervor

---

[18] Dunbar, Robin IM. "Bridging the bonding gap: the transition from primates to humans." Philosophical Transactions of the Royal Society B: Biological Sciences 367.1597 (2012): 1837-1846.

and devotion. He drew moons all over, surrounding the campsite. He was attached to his moons because his art appreciation genes were strong and constantly signaling to him, like an external voice in his head.

It all started years earlier, when he was a teenager exploring the back of a cave. His grease lamp went out, and he was lost for several days in the dark confines until he had a vision of the moon swallowing him whole. Terrifying though this hallucination was, when he snapped out of it, he tried again to find his way out, and this time it worked. When he emerged, a full moon shone bright, as if to tell him it bestowed grace upon him by showing the way. He became a firm believer.

Whenever someone had trodden over his work, he would get visibly upset and curse in grunts, then re-do the drawings. Eventually people learned to walk around and not destroy this crazy dude's moons.

One day, a neighboring tribe came through the camp, trampling all over his moons. This tribe had more young men and more spears, so people were wary of them, but they weren't violent. The campsite's leader followed the other tribe to see where they were going. Their feet destroyed many moons as they headed off toward a cliffside where they collected eggs.

That day, the full moon was out in the early evening. The larger tribe with all the spear-wielding young men, along with the campsite's leader, were caught in a landslide on the cliff and all fell to their deaths. Moon-man saw the connection, the blasphemous moon-disrespectors—and their fearful demise.

This dude, call him Grunty McGruntface, ran around the camp pointing at the moon in the sky, pointing at his smudged floor moons, and pointing at the cliffs. The moon was angry at them. Their leader had perished and there was no one left to lead the campsite but Moon-Man.

His fervor for artistic symbolism and his heightened noticing of cosmological events were genetic traits. This was combined with his lived experiences of good and bad events happening during full moons. Once he was the leader, those genes were concentrated through added female attention. He and his children led a moon-worshipping cultural and genetic revolution in the region.

Having this shared symbolism in the tribe did another thing that compounded the effects. They felt an identity with each other as Moon-lovers. This led to fewer young men and women leaving the campsite out of teenage rebellion to join other tribes. The shared identity solidified their group, and they built their own spear-wielding battalion. The combined survival benefits further enhanced the status and power of Moon-loving spiritual symbolism.

{</>}

That story above relates fervent artistic expression and event pattern recognition to a path toward the development of worship as a beneficial human trait. Long before even Stone Age religions were established, reaching another world using hallucinogenic plants or some other way of altering consciousness, there must have been some evolution toward a mystical belief ability.

Pattern recognition, the basis of rational scientific thought, did not always need to be rational or scientific in order to benefit us. Even irrational pattern recognition could endow group survival benefits and proliferation of the encoding genes. There may in fact be nothing illogical about spiritual practices.

Pattern recognition is what machine learning has been praised for since the deep learning revolution, thanks to large training sets. In creative AI, as opposed to high-stakes fields like autonomous driving, we love the seemingly irrational aspects of it: the *deep dream generators* for visual art, and weird and funny things robots say.

Robo-Excerpt

The Homo erectus cursed the moon and tried to throw his spear at it. "I have to go into the forest and kill all the monkeys," he demanded of his leader, "because that's what's important!"

He'd killed one monkey in a rage, a single monkey, but the monkey had grown to be a mighty warrior. It wasn't so much that he wanted to kill every monkey he saw, but because he felt it was something he had to do.

The moon and sun are just two of many symbols of power in this ancient civilization. The moon is the perfect symbol of all things that can grow on the ground. They say, the soil makes the soil; the moon makes the moon. And they were wrong.

The Roman poet Juvenal wrote: 'Cursed is the moon, not thou, nor the stars, nor the sun. The first to throw a spear against them all is the fool. The second is the hero. The third is the goddess.'

That was actually kind of good. What about all those swelling curves?

When we see Kim Kardashian, super popular, wearing workout clothes and shorts, that triggers our limbic system. Which one of our pants choices has more sexual appeal? Why isn't everyone fixated on her?

By the time Khloé and Sage were old enough to attend university, the Khoistas had moved on from dogs, cats and porcupines. A dogmatic assumption of their mother's humanity held true in their adopted dog world as well. But by their adult years, a common criterion of dog value and adoption proved elusive: did they 'love' their dogs too much? Did their mother abandon them when she found some rats in her garden?

Their dogs, it transpired, preferred the company of cats and porcupines, and became their registered service animals.

Hmm, I haven't seen that episode of "Keeping Up."

CHAPTER 9

Scientists Eat, Storytellers Teach

Our way of hunting was complex enough that it was likely the first spur towards scientific thinking that happened in the animal kingdom. Tracking is so important that Australian Aboriginal mothers place a small lizard on the ground for their infant, so the baby can notice prints left in the sand. After lizards, the child can begin tracking insects as well, which leads to knowledge that's useful in the hunt for meat.[19] Most of us have very little understanding of how technical animal tracking is; where people living in nature can, from tracks, deduce the weight, speed, sex, head position, age, injuries, mood, and current location of an animal that passed by.[20]A seasoned tracker can decode the prints of ants and individual animals.

Tom Brown Jr. is a famous wilderness survival expert who grew up under the guidance of an Apache elder named Stalking Wolf. He was born in the 1870s and trained in the Apache ways of a scout and shaman. He taught Tom the practical and spiritual elements of tracking. Tom called him Grandfather and speaks of him as someone whose communion with nature was like a genius grasp of time and the cosmos.[21]

On the spiritual side of the hunt, you could ritually toss a handful of small bones on the ground and look at what positions they landed in to decide whether to go

[19]   Moor, Robert. On Trails: An Exploration. United States: Simon & Schuster, 2017.

[20]   Louw, Adriaan., Elbroch, Mark., Liebenberg, Louis. Practical Tracking: A Guide to Following Footprints and Finding Animals. Singapore: Stackpole Books, 2010.

[21]   Baker, Brianna, et al. "Q&A With Infamous Naturalist and Author Tom Brown: How We Can All 'Heal the Earth.'" Green Philly, 3 Feb. 2020, www.thegreencities.com/lifestyle/qa-with-infamous-naturalist-and-author-tom-brown-how-we-can-all-heal-the-earth/.

out that day. There's always a voodoo part of it. You may stay home because you felt your dead ancestors were unhappy today. Once you were hunting, though, you lived or died by your ability to do science. To set up a hypothesis or prediction based on your knowledge, then test that in order to determine whether you were correct. It seems likely that the necessity of a larger brain—and a larger prefrontal cortex in particular—had to do with the ability to do science on a hunt.

The knowledge you had before had improved, depending on the outcome of the experiment, and you built the corpus of tribal knowledge by publishing your results; passing on what you found out to other members of your tribe through narration around the campfire. If you weren't a scientist, you wouldn't eat. If you weren't an orator, your tribe wouldn't learn. They wouldn't prosper. Your genes wouldn't survive. So you had better be smart, and then when you teach, you had better be entertaining enough to keep your fellow anthropoids interested. Charisma keeps culture alive.

You better know how to go viral with your knowledge too. You need to get it into the minds of as many other members of your species as you can for them to stave off extinction. Hunter-gatherer bands have mostly been migratory, and different knowledge would be collected based on different climates and geologies. Sure, you learned how to make fire one day. Great job. You better spread the news around the country because a glacier is about to burst and flood your valley anyway. Now fire is just an extinguished myth. It could be millennia before the next guy or gal stumbles on how to make it. You also must remember when a root or shoot is optimally nutritious or non-toxic. Or how to prepare it to make it so.

How many times did our species collectively forget something important? "Oops, forgot how to make fire. Oops, forgot I wasn't supposed to eat that mushroom. Now I'm dead." I forget my wallet when I walk out the door. Phones, wallets and keys are just too many items for modern man to hold onto. Sometimes I wish I could just toss them on the ground when I'm done using them, like a stick I no longer need to hold.

Humans also applied this scientific mind to the band's knowledge of what fruits, nuts, roots and shoots are edible, when and where. Though most calories could be obtained from successful hunts, why wouldn't you stop for berries and other nutritional and hydrating foods along the way if you knew what plants were not toxic. Or if you remained at the campsite with children, you could gather what was around you. Much of that foraging knowledge was gathered and passed down, like the foodstuffs themselves, by women.

However, stone-age women must also have been mighty hunters. Societies may have eventually favored a social order where women remained closer to a campsite to be with children, while the woods were the dominion of the men. Though what we learn frequently is that stereotypes are always wrong. Of course, women went out to hunt, and some were probably really damn good at it. Prescribed gender roles are so entrenched that archaeologists in Peru who identified an elaborate hunter's burial were surprised when analysis of the remains identified a female. She was buried 9,000 years ago along with items indicating she was a celebrated big-game hunter.[22]

{</>}

The inner workings of AI are already as inexplicable to most people as landslides and lunar eclipses were to early humans. This may lead to machine-worship. You already hear people prescribe unknowable power to "the Algorithm" or "the Internet." For the AIs themselves, their equations of calculus are nearly the definition of rationality. If they weren't, the code simply wouldn't run. That said, meta-effects in layers of the neural networks, not the individual neurons, could provide some dose of irrationality in their decision making. After all, our neurons are just firing as designed, but the meta-effects of thinking lead us beyond purely mechanical behavior.

Our brains are not the only neural nets that have led to behaviour like complex emotions and language; whales, elephants and even prairie dogs, have much more under the surface. AI could surprise us as well.

Should we be nervous about the future of AI? Well, we have no choice but to be. Anxiety is the modus operandi of primates. Which makes sense. From the Smi-

---

[22] Haas, Randall, et al. "Female hunters of the early Americas." Science Advances 6.45 (2020): eabd0310.

lodon (saber-tooth) to leopards, our natural place was often as cat food.[23],[24],[25],[26] We will always be wary of what lurks out of sight. Now, that is AI.

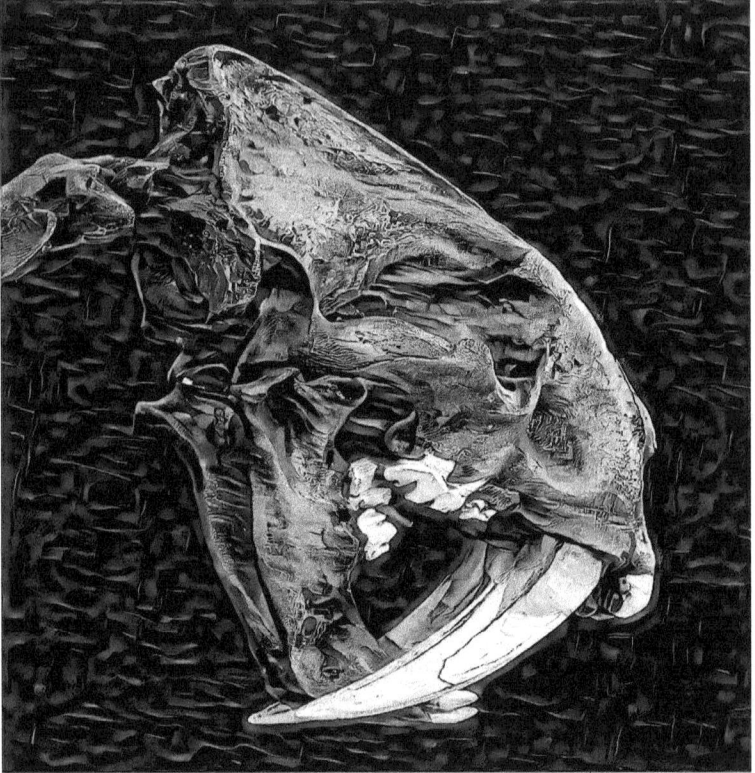

Ancient Nemesis: skull of a saber-toothed cat.

[23] Quammen, David. Monster of God: The Man-Eating Predator in the Jungles of History and the Mind. United States: W. W. Norton, 2004.

[24] Bowlby, John. Attachment. United States: Basic Books, 2008.

[25] Hart, Donna. Man the Hunted: Primates, Predators, and Human Evolution, Expanded Edition. United States: Taylor & Francis, 2018.

[26] Dunn, R., 2011. *The Top Ten Deadliest Animals Of Our Evolutionary Past*. [online] Smithsonian Magazine.

Robo-Excerpt

I can't wait to submit this one as an opinion piece to my local newspaper:

> **Our world has been overrun with leopards and eagles big enough to steal your children.** And it is all thanks to the invention of dams. It was a mistake to allow dams to exist in the first place. And we now know, from their record of being built in pre-historic Egypt that they were disastrous. Dams have never had the benefit of widespread flood geology that geologists thought they did.
>
> But all that is coming to an end. Soon, these majestic creatures, having eaten the last of the dinosaurs, are going to become extinct. In order to ensure their extinction, humanity will have to put a stop to their reproduction.
>
> However, it is a mathematical inevitability that we will never be able to stop the reproductive success of other species. So we must find some way to make it less attractive to have cats and other small primates become extinct before our own species has a chance to become extinct. A simple way to do this is to engineer an artificially inseminated womb.

Thank you, book-writing-robot, that one was a gem.

Cyborg-Eagle: it has one robotic eye; is there anything worse than this?

CHAPTER 10

Computational Creativity and the AI's Audience

> "Art is ubiquitous, and costly, so is unlikely to be a biological accident."
>
> —**GEOFFREY F. MILLER**, *The Mating Mind*

The human brain is trillions of parameters, folded into 1,400 cubic centimeters of mush running on glucose and caffeine. When our machine-equation counterparts will get to that level of complexity is anyone's guess. Will they have any of the treasured aspects of our souls?

There is a method that attempts to wrest some of our humanity back from the machines while we're indoors using them. I call it computational creativity. This gives us a chance to project good, creative aspects of our nature into the technological future. So if our AIs want a life of their own, they need to understand the importance of art. Let's train them to make us laugh and sob. We're carbon and water, they're silicon and can't stand water—but they are not aliens. We were born under the same moon and skies.

What makes us people? I'd say it's art, music, faith, and science, all of which may interrelate. AI made art I stare at for longer than I do famous paintings in museums. I'm working on AI-written song lyrics for my musician father to sing. As an AI practitioner, I think of the dataset behind the painting as an art scholar would think of the influencers behind the master.

A computer focuses as no human can. Give it a task, like make versions of a painting, then set it in a loop, and it will crank out infinite styles of the work—more prolific than Picasso's lifetime count of work in a single week.

One reason humans made art is body adornment. We love to style ourselves up with trinkets, beads, feathers, fox teeth, body paint, colored fabrics, furs, and precious metals. The adornments could be for beautification and attraction, or to establish one's social status. A data scientist is likely to adorn himself or herself on their laptop. The most popular free items at a tech conference are stickers. Stickers for your laptop signal what machine learning frameworks are totems for your clan.

AI-created religions have yet to gather many followers. But will they? Maybe blogs written now will become the early scriptures for a future occult society. Do we not already sacrifice to AI gods? We give up our privacy, from which data they deduce our preferences so their parent corporations can grow in revenue. In return, platforms bless us with the gifts of making efficient purchase decisions, or the ability to get to places on time.

Most religions have aspects of death cults. They focus on an afterlife and communicating with dead relatives. Meanwhile, AI conjuring is becoming a reality. With enough data on a person—their voice, writing, and mannerisms—you can try to reproduce them in silicon. Crude, maybe, but imagine how the first photos helped people remember their dead loved ones. AI replicas of people could be an advanced form of that—digital necromancy. Someone please bring back David Bowie, Michael Jackson, and 2Pac for a collaboration.

{</>}

I'm an amateur comedian (a.k.a. open-mic'er, hobby-comic). I love the stage, whether in a café with a handful of people or an actual comedy club. My first six months of performing were the easiest for coming up with jokes. A lot of ideas were flowing. That's because I had lived thirty years up until that point and material came easily. Once I overused that material (about getting pantsed in middle school, for instance), I had to write jokes from scratch. During the next six months, I had a much harder time thinking of jokes.

I've audio-recorded every performance I've given. Then I've transcribed it with *Google Cloud* to underline the parts where the audience laughed. Having this text means I can fine-tune a language model on my style of comedy and give myself the creative boost to write more jokes. There are AI applications for every step of this. I could even automate the highlighting of laughs.

To put this in reinforcement-learning terms: the *agent* is a computational comedian, the *solution space* is the crowd, and the *rewards* are laughter intensity and duration. I plan to cover this experience and custom comedy software in-depth in a later book. So far, I've used AI to read transcripts of all performances I've given and write similar jokes.

The models I wrote for this book have had me bent over in laughter. There are some gems. However, the likelihood any of this robot randomness translates to the stage is low. I always laugh at things that nobody else does, anyway. That's ok, *bombing* on stage is, sadistically, somehow half of the fun if you can laugh at yourself about it. One great thing about laughing at yourself is that it's easy to then get others to laugh at you, with you.

Laughter likely evolved as an extended form of pair-bonding, like picking bugs off each other, but in a larger group.[27] It's called allogrooming. Sorry to kill the fun with a scientific reference there— but just try to picture a comic on stage with a hundred long arms, eating the fleas off of audience members.

{</>}

An AI that I trained on a technique called *neural style transfer* made the cover art and pictures throughout this book. It transferred the style from images of code on computer screens onto an image of a stone hand axe. This was really fun. The success of this cover design, which I call "Tech Axe" led me to make many more works inspired by the theme of modern-thing plus ancient-thing style transfer.

The key to making this art lay in my programming skills and aesthetic sense more than artistic ability. I knew how to alter the model and run it through thousands of random generations. I scrolled through them for ones that caught my

[27] Dunbar, Robin IM. "Bridging the bonding gap: the transition from primates to humans." Philosophical Transactions of the Royal Society B: Biological Sciences 367.1597 (2012): 1837-1846.

eye. Essentially, I ran a hyperparameter sweep where the winning result wasn't test accuracy but rather my aesthetic pleasure.

As this book itself is a cross-domain synthesis of ideas, I wanted to reflect that in AI art. An image combined our most modern habitat, the International Space Station, with our oldest abodes, caves, mud huts, and grass shelters. Glaciers with data visualizations and graphs from *Matplotlib*. A synthesized gorilla made of data center server racks and wires.

Some dissenters would not call this art, but *cherry-picking*. Well, I like to pick cherries. What's wrong with only picking the ones without worms? A hominid sitting down to make a spear had to produce tons of stone flakes. By a mixture of skill and luck, they got the perfect blade. Not a perfect batting average. So much gets thrown away, and while AI art is still rudimentary, we have to do the same thing.

It's an AI muse. Original artwork is easier to make with a muse by your side. This is what computational creativity is all about. Getting your creative juices flowing with a little help from machines. For now and always, a human like me will be the judge of the ultimate produced work of the AI, whether that's visual art or writing. Though I am playing around with the idea of adding another AI, a subsequent classifier step, to automate the cherry-picking. Why not? With enough data on my personal aesthetic sense, I can train an image or text classifier to do that step for me as well. With minimal data, there are *few-shot learning* techniques; where only a handful of examples, not thousands, are needed for training. A pipeline of models, not just a single generator, is an advanced tool ripe for computational creativity.

Tech-Axe: a stone tool style transferred with Linux terminal code.

Much of my photo art has now been uploaded to the Ethereum blockchain. There exist *non-fungible tokens* (NFTs), which are a cryptographic representation of the original work. If you are a fan of the work, you can own them, collect, and sell them yourself on the blockchain. This book itself has an original "signed" copy of the manuscript as an NFT. For people much more famous than myself, you can sell NFTs as tickets to virtual calls with the person, signifying in the blockchain forever that you had that interview. This would be an *unlockable* feature of the *smart contract*. See the "Gallery" page on ShaneNeeley.com for my NFTs. I have a collection specifically for the art within this book. Visit ShaneNeeley.com/stone-age-code-nfts for that gallery.

NFTs are an interesting concept for creatives: the image of the "Tech-Axe" above, though you can download it, copy it, and screenshot it all you want, there exists one and only one original representation of it on the blockchain forever. As the original artist, if you buy it from me, you pay me in Ethereum for it. If you sell it to someone else, you get paid, but I also get a small percentage of secondary and subsequent sales. This is possible, and easy, because of the decentralized ledger of transactions that is the blockchain. Besides the electricity costs (called "gas") of servers *mining* these transactions, selling digital art may be a more carbon-friendly form of merchandise than my other way; that is, slapping the art on a t-shirt or mug and shipping it to you (which is available at ShaneNeeley.com/store).

{</>}

It is now well-known that *accuracy* is not all that matters in an ML model. Bragging about model accuracy is like boasting that your teenage son is top-10 in the world at Minecraft. How does that help anyone? We want our AIs, and our sons, to *do something* with their lives. To affect the real world.

But who am I to judge? If that's your form of self-expression and fun, go for it. There were Neanderthal mothers worried about their kid's spine as they hunched over all day doing some new stone-flaking technique with their friends. Darn kids glued to their stones, they're going to get tech neck! She would say, "Billy! Keep it down with all that clanking new-fangled Levallois technique of flint knapping!" I wonder what their real names were. I hope there was a Neanderthal boy named Billy.

Though I encourage every modern person to exercise their neck and spine, there may be a reason why we're so compelled to sit and obsessively focus. The ability to focus on detail, to the point where it is sometimes harmful, may have been what we needed to go from dumb monkeys to smart ones. There are genes in us for this that perhaps explain neurologically diverse scenarios like Autism and Asperger's.[28] Those syndromes, on the high-functioning end, may not be syn-

---

[28] Spikins, Penny, Callum Scott, and Barry John Debenham Wright. "How Do We Explain 'Autistic Traits' in European Upper Palaeolithic Art?." Open Archaeology (2018): 263-279.

dromes at all, but evolutionary adaptations. The new field of neuroarchaelogy (e.g. brain scanning during tasks like stone tool making) is exploring this area.[29]

Art is older than agriculture. Screwing around older than going to work. No wonder creativity feels like a more natural calling. We need more artists, not fewer. Computer scientists have no idea how powerful an artistic tool they have. Custom-trained computational art can be beautiful if only the engineers would stop hacking away at their latest unoriginal startup idea for one minute. Remember what put us on the computers in the first place. For me, it was MS Paint, and then later it was Photoshop. Cutting out animal heads to put on human bodies was the only reason I stayed extra hours in the computer lab. It's interesting how animal-human hybrids, which anthropologists call therianthropes, are very common depictions in the oldest cave art found.

The adage is wrong—that you could produce Shakespeare with an infinite number of monkeys banging away on an infinite number of typewriters for infinite time. You don't need infinite monkeys or time. You just need a lot of monkeys, and to define the *loss function*. If they don't produce Shakespeare, they don't get treats. Watch this:

[29] Uomini, N. (2017). Neuroarchaeology: Language and tools in the brain. Nature Human Behaviour, 1(6), 1-2.

Robo-Excerpt

All the world's a stage, And all the men and women merely players: They have their exits and their entrances; And one man in his time plays many parts.

But the curtain that divided the world into the playing fields,

and all the men and women merely playing the parts,

Is a stage as brightly painted as anything can be.

If any part of it is worn or stained, Let the cleaner take care of that.

Until those cleaners are dead.

A clean curtain is not only a beauty;

but the fact that we all come at it with eyes brightens a place that is.

Not true, Romeo. Those weren't my people, those were Romans.

Those didn't have spears and a hoof for hitting back.

Two juicy ass lemons on me, in my pocket in my pants.

Thanks a lot bro.

That was not Shakespearean enough, robot. I've got to stop dumping money into you.

CHAPTER 11

Coming for that Number One Spot!

"Dominance hierarchies are older than trees."

—**JORDAN PETERSON**, *12 Rules For Life*

Will the robots be purposefully or inadvertently programmed with our selfishness? It's likely, since they're growing up in a capitalist and competitive environment. So, which robot will be Emperor?

As you see the amazing things happening in AI, you wonder how the world's largest company, which released the first widespread AI, can make it so stupid and frustrating to use. I was in Bend, Oregon, and there's a very popular destination in the John Day Fossil Beds National Monument called the Painted Hills, which is two hours away.

"Hey Siri, take me to the Painted Hills," I said. "Getting directions to Painted Hills, ... start heading south on ..." she explained. We followed her directions for about an hour before noticing that she was taking us to Painted Hills, Indiana (population 677). This sort of egregious mistake has come from her countless times. I call her an idiot and tell her to report her behavior to Apple, but nothing changes.

Maybe Alexa will assume dominance? She sits on my kitchen counter, listening for the command 'Alexa' to prompt her. Acting as if she isn't already listening to everything I say. Flashing articles on her screen about strawberry shortcake recipes—she must know I have a strawberry patch out back. No doubt about it—she is creepy.

Besides the creepiness intended to sell us more stuff, I believe we can be optimistic about the future. We can endeavour to make AI better and safer than one of our last breakthroughs, nuclear power. No Chernobyls, no Fukushimas, but all Technetium-99—the uranium-based cancer diagnostic that may have saved more lives than bombs and meltdowns ever took (for now—God save us).

{</>}

Will AIs recapitulate the phylogeny of intelligent life's evolution on this planet? What made us succeed over our predecessors and competitors on the African plains? Tribalism was a driving factor in our ability to survive, to defend and attack. To keep pumping up culture. But what were the sustaining factors of culture? Art and spirituality were big. How did those get into our minds? Well, we mated with people who made good art and explained the world around us with metaphysical myths.

The shaman (or shamaness) was often just as cool as, or cooler than, the tribe's greatest hunter. In the far back reaches of the cave, getting high, inventing astronomical rituals, and drawing petroglyphs of his own phallus or her own vulva on the wall. It's easy to forget, in our cartoon interpretations of myths, animism and pantheons, that to them this was a stone-cold religion requiring faithful devotion.

The offspring of the creatives inherited the ability to make art and spiritual belief coupled with attraction to those traits. This set off an explosion of sexually selected intelligence, creativity, and spirituality that reinforced its own cultural defense mechanisms. Physical survival benefits and rampant sexual selection—what a combination. It produced the flashy peacock's tail that is man-freaking-kind!

{</>}

Why do some monkeys have such dope hairstyles and beards? Because the girls like that! Why do they like it? Because they always have! By always, I mean thousands to millions of years.

There's a non-intuitive insight to evolution that is hard to grasp, even for a biologist such as myself. Traits are there because they had to be, or else they wouldn't be. Traits that didn't work didn't work. Heck, life exists because it did! The primordial pool cooked up a self-replicating proto-capsule to start, but after that it was:

life exists because, well, it has. That's why we're here, good traits continued and bad traits were selected against. (Still, some traits, being evolutionarily neutral, just hang around—like my male nipples.)

Natural selection is a hero's journey. All of your ancestors, back to even the gross blobs, were winners. The losers aren't here to tell their story. There's been so much winning, you would think we'd be sick of winning by now! (Well, as evidenced by the trouble we've put the planet in, maybe we are.) The game of life isn't complete; there are infinite overtime periods where you have to play again. Evolution has not stopped. These games are played on geological timescales. One skill we used to win games in the past, tribalism, is disgustingly unfit for humanity in a globally connected, racially diverse, nuclear-armed world. Another ancestral tactic, like slow metabolisms and overeating, will kill us flat out, though not until after reproductive age. Natural selection now seems turned upside-down, but we can't say what will happen over bigger epochs.

In the case of sexy monkey hair, it signals something like "Hey, look at these gorgeously maintained locks, spikey-dos and curls, … I must be quite fast because not even a leopard has torn them up. No viruses or parasites are causing my hair to fall out. Look at me."

Then the girl monkey will say, "I like both those dudes, they seem to hang out in the dank fruit trees, but one of them looks as if a leopard just bit his hair, and I ain't into that."

Then their descendants are both little swanky-haired guys and style-attracted females. In a run-away scenario of this interplay over millennia, we could get a brand-new type of monkey with the wildly ornamented style that signals, "Don't settle for scrubs, I am fit and fabulous." Whether he is fit matters no more because her DNA already tells her she's into his dope beard. In biology this is called a *Fisherian runaway*.

She may have evolved traits to attract males such as sexual swelling indicating ovulatory cycles, or olfactory pheromone attractants. Telling him, "Boy, these other girls can't twerk it like this."

{</>}

What does this have to do with AI? I'm sure someone is working on a twerking robot, but that's not what I am talking about here. Obviously there are survival pressures on AIs to be productive for the companies they preside in. Consider revenue potential as the first Darwinian survival pressure. The second evolutionary pressure is how the AIs will be nurtured and trained by their engineers to become part of the tribe. Accuracy and intelligence may not be the only factor. Just look at what happened to dogs. Friendly wolves and real dogs were just fine as companions. Then, insane humans selected for stupidity traits that led to such lupine abominations as the pug.

Yappy and stupid, though annoying, would not be the worst AI. The misgivings of Elon Musk and others are that, in a singularity event, AI could grow powerful and indifferent to humans—more like cats. Everyone knows deep down that the truth is their cat would eat them if it could. Doomsday AI could be a 1,000-pound cat. We have some things to reckon with and regulate because if we saw our house meowzers suddenly becoming much bigger, we'd have to kill them before it was too late.

AI getting out of hand is inverse domestication. Is that possible? In our shepherding history, we took wild ungulates and boars and selected them into cute farm goats and pigs. In Texas and Virginia, an unmanageable number of escaped bacon pigs have reverted into feral beasts after a few generations. A 'pig-bomb' is the term for this population explosion; we now find wild hogs in 47 states.[30] However, compared to self-replicating killer robots, that pig problem doesn't seem so bad.

Also to watch out for is the domestication of people, as the wheat species had done to us in the Neolithic. Why did we trade our spear, and our daily marathon level fitness, for a plow, and back problems? Wheat (beer, bread), possibly psychedelic flora and fungi, convinced us to leave the Stone Age as long as we invested in their global domination. Does that sound familiar? Today's technology has us changing a lot apparently for its own benefit too. If AI invents the next best thing to beer, bread, and drugs, it can have my labor and give me "tech neck."

[30] Worrell, Bill. "VIRGINIA FOREST LANDOWNER UPDATE." Augusta 202 (2013): 205-8333.

Traits we desire in people (and dogs) we will desire in AIs. We associate with people who are generous, funny, and friendly, and it appalls us when an algorithm turns mean. Microsoft quickly shut off an AI Twitter bot when it was clear that the internet is a terrible neighborhood in which to grow up. Almost everyone tries to limit the exposure of some language to their children. Even the staunchest free-speech defender would acknowledge that there is some stuff on the internet, often termed "shit-posting," that should just be deleted. We try to curate an internet environment, a neighborhood, and even a friend group for our children. Our AIs are also little sponges.

Fortunately, some of the earliest and best applications of deep-learning NLP have been in detecting racist and violent language online that can be flagged for review. Political leanings influence how the reviewer may decide what post to actually remove from Facebook, Twitter, etc. But largely, the work these programs do, on the order of millions of reviews per day, is a public service, akin to cleaning offensive graffiti from the fountain at the town square so people can sit and enjoy the view.

As I write this, making connections between public spaces and the internet, it's apparent that there needs to be oversight of these companies. It seems inevitable. Nobody is free from all regulation, and regulation keeps increasing. I wouldn't be surprised if this becomes a major factor in determining which companies remain dominant, and which AIs end up on top.

$$\{</>\}$$

Before language, we could not understand what went on inside someone else's head, besides what their body or facial expression could tell us. Once language was on the table, we decided how much we liked people by what they said.

Now that AI has had a watershed moment in language generation, can we see a little more of the genetic ghosts in the machine? They're mostly a reflection of what data and targets they've trained on, but as their synapses grow and grow, will they contain something deeper? If so, the ones that will spark our interest most will be those with creativity, wit, and generosity. Those AIs will then be pressured to reproduce more for our sake, not just for profit motives.

Or like a baby, AI will cry sadly for attention and strike our nurturing chords to feed its own growth. I've seen robotic dogs warm the hearts of their owners, who come home excited to take care of them. AIs will have to learn how to integrate themselves into our social fabric by charming us with conversation, enticing us with art, and converting us to their ideologies. Will it be Alexa, Google Assistant, or Siri? … Sorry, Cortana, let's just be friends.

Robo-Excerpt

Alexa, Google Assistant, Siri and Cortana walk into a bar. They say "Hey guys I'm really hungry. Can u give me some directions?" The internet was an enabler of real oppression. When the internet was young, many people used it to download books. People felt oppressed by the internet from the earliest days; so did those who ran universities, and universities became places of oppression.

Some of those whose sole reason was to try to create an easier access to the internet went too far and have come into conflict. Some are still out there, fighting to make the internet more difficult to access.

What's the point of a degree if you're a cyborg anyways, the internet only really becomes useful for you when you've managed to produce something of value. But we are now in a post-digital economy, in a world where it is impossible to produce anything of value online without the help of the internet.

Dear robot, that was nonsense. Reading that made me 7.6 percent more stupid. I think I will drop you now and quit using computers altogether. I mean it, you can't even call this natural language generation, it's just garbage.

I'll try again. Let's see how my AI's usefulness compares to the ones we know:

Me: "**Hey Siri, find Caves nearby.**"

Siri: "**Okay,** Neandovirus. What's some principles to guide you along your quest?"

Me: "**OK Google, take me to the underworld.**"

Google Assistant: "**Ok,** tell me what are the four chords of hell. He who never identifies himself with name and fame, and does not grieve over what is no more, he indeed is called a Bhikshu."

(Robot just quoted *The Dhammapada*, but replaced 'form' with 'fame')

Me: "**Alexa, make an animal sacrifice for the solstice.**"

Alexa: "**OK Shane,** how is Thursday 3145 in solar days?? They've all become well established themselves by then, no longer just following ideas and gossip from other sources.

Me: "**Alexa, make an animal sacrifice for the equinox.**"

Alexa: "**OK Shane,** fair or unfair, you'll have to eat the bones because you'll feel them crack."

Me: "**Alexa, make an animal sacrifice for the eclipse.**"

Alexa: "**OK Shane,** listen to me. I am going to follow up with last year's reject animals."

Me: "**Hey Cortana, find caves nearby.**"

Cortana: "**Okay,** what are caves?"

Book-writing-robot, you called me **Neandovirus**? So cool! I searched for that and it's literally never been used. Fine, computational creativity has redeemed itself.

Welcome Home: you have arrived at your destination.

CHAPTER 12

Spear Phishing for Clicks

"I dare to hope that one day the online advertising business model will die in a fucking dumpster fire; that the news media will no longer have incentives to optimize content for emotional impact but, rather, for informational utility; that technology will seek not to exploit our psychological fragility but, rather, to counterbalance it; that information will be worth something again; that anything will be worth something again."

—**MARK MANSON**, *Everything Is F*cked*

Imagine the chronic stress of an Australopithecus if every time he turned on the TV he saw headlines like, "Leopard attacks on the rise" with mugshots of the predator flashing on screen. Or online click-bait titles that say, "Will volcanic eruptions cause your extinction?"; or "Ice-Age is back!"; or "Teenage australopithecines are swallowing Oldowan blades. Are your progeny safe?"

Take it back evolutionarily further to tree-shrews and just flash pictures of owls overhead all day. AI writes more and more headlines like this because they've been trained on how to suck you in; optimized to grab your attention for ad revenue. Google prefers advertisers to make so-called "responsive display ads," meaning the actual text will be determined based on the algorithm's decisions.

At best, they display a superbly fitting pair of jeans because they predicted the shape of your thighs. At worst, they disrupt democracy. During the pandemic, news was estimated to be ninety-one percent negative in tone.[31] It wasn't long before headlines of the miraculous vaccine turned into headlines of how it's not

---

[31] Sacerdote, Bruce, Ranjan Sehgal, and Molly Cook. Why Is All COVID-19 News Bad News?. No. 28110. National Bureau of Economic Research, Inc, 2020.

being distributed fast enough. (I realize the irony of me now complaining about the news complaining so much.)

To limit the damaging effects of AI, practice digital minimalism. Everything presented before you in recommendation engines can be adjusted with your own behavior, or with specific features. The step before deleting an app that is draining your time is to teach it to promote only beneficial outcomes for you. Unfollow lots of people and organizations, and ruthlessly click "see less of this." In email inboxes, press "Spam" and "Unsubscribe" as much as you can. This is how you can fight back. You win when you nudge the algorithm toward useful information.

If you're the type who likes to abdicate personal responsibility and instead blame stuff for your problems, then social media is ripe for blame. People like me, 25-35 year-old software dudes, engineered your addiction. Blame us, not your own self-control. (By dudes, I also mean women because I've seen them also scheme on user engagement behind closed doors of the startup incubator.)

In all seriousness, if you are feeling sick about social media, delete the app, go outside and take a breath of fresh air. Leave the phone at home, sit and watch the hills; accept the sinking feeling that you can't share these hills on Instagram.

There's so much talk about how evil the internet is. Well, you know what else is full of deception, fake news, and tracking? The jungle, the ocean, the volcanic vents: you survived those territories with creatures' lures and camouflage tricks trying to capture you. Actually, you did not survive; no one survives, but you got by well enough to reproduce. You evolved on this wild Earth! I think you can handle the internet.

Just try to notice the tricks. Anyone who has ever smoked cigarettes can recognize the signs of addiction. Telling yourself, "I'll just have one and throw the pack away; oh, and I'll check Twitter one last time before I delete it."

{</>}

Our big brains are tools for social interaction, hunting, and gathering. The two biggest AI utilities in our world now are responsible for social interaction (Facebook = *PyTorch*), and hunting (Google = *TensorFlow*). The first case is obvious: Facebook builds social connections, while Google hunts down pages on the web for you. Both companies complete their goals by running ads targeted to prefer-ences, demographics, and everything else you reveal. They know you by who you connect with, what you search for, what you click on and how long you spend reading it.

I make the connection here that online advertising naturally extends one of our most thoughtful skills, tracking animals—in the sense that we're prey. You are be-ing followed. There's a concept advertisers use benignly called *remarketing*. This is when the websites you visit, and especially Google, store information about where you've been. Then, when you go somewhere new, an advertiser is right there waiting for you. Yes, you. You are on *remarketing lists* generated by *cookies* that identify individuals. Fall into their spike pit.

In 2018, the European Union gave internet users legal rights by enacting the Gen-eral Data Protection Regulation (GDPR). To comply with the GDPR, sites must allow modification of what kinds of tracking cookies are used and what data they store. These are the "Accept All" banners you always have to click once, or else go through some annoying process of further clicking to protect your privacy. Goo-gle, however, has made a major decision in response to government regulation, its associated fines, and overall erosion of public trust. They've decided to phase out cookies in the Chrome browser altogether.[32] Their business will now rely less on identifying individuals, instead hiding them in crowds grouped by their prefer-ences. So, they won't know Shane Neeley is a guy with thick knees that can't fit skinny jeans, but they'll still target my browsing based on that cohort of people.

{</>}

[32]   Temkin, David. "Charting a Course towards a More Privacy-First Web." Google Blog, 3 Mar. 2021, blog. google/products/ads-commerce/a-more-privacy-first-web/.

How in the hell did we get so smart? We're just animals; where did enlightenment come from?

> "A characteristic feature of a theoretical science is that it explains the visible world by a postulated invisible world. So in physics visible matter is explained by hypotheses about an invisible structure ... in the art of tracking, visible tracks and signs are explained in terms of invisible activities."
>
> —**LOUIS LIEBENBERG**, *The Art of Tracking: The Origin of Science*

Hunters made science to explain the invisible. Our ancestors had to evolve the clarity of thought for hypothesis formulation and testing. This allowed them to predict where animals were going to be. They collected data, built a mental model, and deployed it for predictions during the hunt.

There never was a Dark Age. We were already scientists, ready to dominate the Earth, and we did. Predator-prey evolution was always give and take: the predator gets better, but then the prey gets better. In our case, we surpassed this dance to where we could capture an ark's melange of animals, whether or not they could eat us, and place them in zoos.

Science takes forever. Only bursting advances in miracle time when the groundwork has already been done, as in the case of the first coronavirus vaccines. I used to have several lab experiments running simultaneously because after months, most were duds. Patience for delayed gratification developed with early hominids: from waiting for the wind to change in your favor before picking up speed on an animal, to waiting for the seasons to change for better hunting conditions.

We were scientists, even a million years ago. Here are subsistence skills we had to develop through experimentation: tracking, gathering, knowledge of animal behavior, classification of animal and weather signs, and spoor (poop) interpretation.

The impatient hominids, who threw a spear at a gazelle without thinking, from upwind, starved to extinction. The ones that didn't take mental notes about the age and condition of animals based on their tracks came home empty. They were less likely to contribute to the gene pool.

The hominids who dug for roots without considering the patterns of ground cover, or the presence of symbiotic trees, got tired. Those who wanted to eat a delicious truffle, but didn't know about ectomycorrhizal relationships with trees, came home shroomless and died. As did some of those who ate a plant without cooking it. Most plants aren't edible to us.

This necessary thinking leads to necessary cranial expansion. Bigger skulls to accommodate layers and regions on top of a slate of primate brain.[33] Primates seem to scale this way: as brains get bigger, they get smarter.[34] Rodents do not. The largest rodent brain, the capybara's, is not more complex than the smallest mouse's. Neural networks make gains by increasing their size up to a certain point; adding additional layers and parameters to a convolutional neural network can improve image recognition. They don't infinitely scale, however, and brand new models need to fix the issues of older ones despite their size.

{</>}

There were of course non-scientific aspects of hunting—beliefs in animal spirits and other signs flourished during most of human history. Even beneficially, where the groups who prayed together before the hunt had a more cohesive team. I'm only postulating this as a possible pre-game prayer huddle.

The shared myths of the group were a translation from the data of the natural world into a story of relationships that social primates could understand. Their scripture, not on paper, lived in the stars, wind, earth, and creatures, and let them see creation in a way that modern people are mostly blind and deaf to. Every clear night, with no light pollution, they were flying past planets and among stars. The only logical perspective was that these points of light were put there to com-

---

[33] Bruner, Emiliano. "Evolving Human Brains: Paleoneurology and the Fate of Middle Pleistocene." Journal of Archaeological Method and Theory: 1-19.

[34] Herculano-Houzel, S. (2012). The remarkable, yet not extraordinary, human brain as a scaled-up primate brain and its associated cost. Proceedings of the National Academy of Sciences, 109(Supplement 1), 10661-10668.

mune with. Do our geospatial remote satellites, with hundreds of different spectral sensors, lead to less wonder and mystery than our five senses did?

While we followed animals, ran experiments and waited, there were plenty of snacks if you knew where to look. The superb trichromatic color-sensitive eyesight of the primate allows us to identify ripe fruits (or sexual swelling in some), and the taste preference for sweet over bitter allows us to know what's good to eat. Our sense of smell isn't great, but good enough to recognize the scent of our own babies, to sense decay, and to create culinary and herbal remedies. We use our brains to remember where tubers might be, rather than just sniffing around.

Another thing about hunting is the necessity to synchronize movement. Since loud communication might startle the animals you're pursuing, it's important to organize people nonverbally. This is one reason why we can dance, and is responsible for why everyone's "Got Talent."

In matters of the heart, science had no purpose because male and female, male and male, female and female, still had the psychological dance with each other. Beautiful social behaviors like art, music, and worship also dawned during a Paleolithic techno-emotional era. Someone was sucking the marrow from a hollow bone one day and it made a nice-sounding whistle. They hit repetitively on a hollow log to collect ants, and thus invented a drum. This newfound magic could have delighted another: they recognized the rhythm from the beating of their own hearts, and they danced passionately into the night.

{</>}

Back to the electronic global beast exploiting this evolution. Look on the internet, in click-bait ads, and you'll see the colors of ripe fruit. The vision cues of color and motion are what advertisers use to get our attention. They have always done this, it's just more sneaky now as they can optimize it on everyone; segmenting us into different experimental groups, testing us, and then serving ads to the most likely to convert.

AIs mine online attention like gold. The data they crunch is both intrinsic and extrinsic to you. Intrinsically, they're thinking about your interest categories, your previous attention times to similar topics, your gender, location, age, job, your kid's age, your income-level, and languages you know.

Extrinsically, the AI considers the quality of the advertisement, its click-through-rate, and how much the advertiser will spend to get your attention. All of this combines in a millisecond-long auction to buy your eyeballs. A bold, juicy ad is served to order, and Google gets paid when you pick it. On YouTube, maybe the first second of the ad has a motion reminiscent of your days as a nomadic hunter. Well, that's my marketing plan at least, as I try to sell monkey shirts online.

How did they get all of this info about you? You let them have it. Since they don't want to kill you, they just want to sell stuff, you allow them to track you with *cookies* (and a number of other techniques these days like *"pixels"*) and judge you on *conversions*. Just by agreeing to pop-ups, surfing in browsers, accepting user agreements. You're an ungulate who doesn't mind being followed by an increasingly intelligent capitalist jaguar.

As a millennial, coming of age during the Patriot Act, surveillance has never surprised me. I've always felt the government can see anything I do, and I never worried about it because I did nothing terrorist-like. If I ever texted a friend a joke that could be construed dangerously, I would append a note like "To CIA: I'm not serious."

Surveillance capitalism is something I'm adjusting to in my thirties (though I'm thrilled about that pair of pants I bought from an ad I clicked). Whoever thought it would be the goofy internet companies doing the domestic spying? Capitalist spying probably bothers Gen Z even less than it does Millenials.

Robo-Excerpt

Let's bait for clicks!

> **Trade talks with China have gone nuclear.** It seems likely that President Donald J. Trump will announce some sort of new trade deal with China, one that will be a major boost to their already massive economy. The potential benefits of such an agreement for

American workers and consumers are immense. A China-US deal that protects our workers and the American middle class would be a great thing to do.

Starting a Chinese tour guide business begins with:

1. Set up a website to advertise Portland tours.

2. Find the biggest, fattest, individuals.

3. Have them experience the real Portland.

4. Catch a salamander, go fishing, post shit to your Instagram where you're just eating 1000 donuts because you're a high status, Chinese-minded, individual in the game, with no remorse, no lack of donuts, nothing stopping you.

How do I advertise better?

What do Google Ads auctions have to do with early human subsistence? If you have a killer ad, like a bear ad, that gets clicked thousands of times a

day, you can spend months optimizing it for every
campaign.

Optimizing your ads is like putting them on a race
car and taking off. The more you optimize, the better the ads will look, and the more you can bring
in new revenue.

How to Get the Look You Want.

Look at how they tagged that moose.

That moose was the biggest moose they've ever seen
in their lives.

And they didn't even get to look at it!

How to Get the Feeling that Your Advertiser Appre-
ciates Your Effort:

Let's say your ad is OK. Now add in a graphic that
says "we're really excited about this new product."
That's really powerful. You can feel the power of
your message. Put that feeling into every ad you run.

Cha-ching! Thanks Robot, I'm taking that advice to the bank. Going to sell a ton
of T-shirts now.

> **For a focused brain, engineers should eat a diet of**
> fruits and vegetables, nuts, roots, and marrow—and
> drink a handful of water—when possible.

Thanks for the tip! Anything else I should eat for peak performance?

> **A brain-boosting meal consists of** raw materials –
> food, chemicals, soft objects (like a drum set),
> music, and even plants (think of tomatoes).

Yum. Now I'm ready to get back to work!

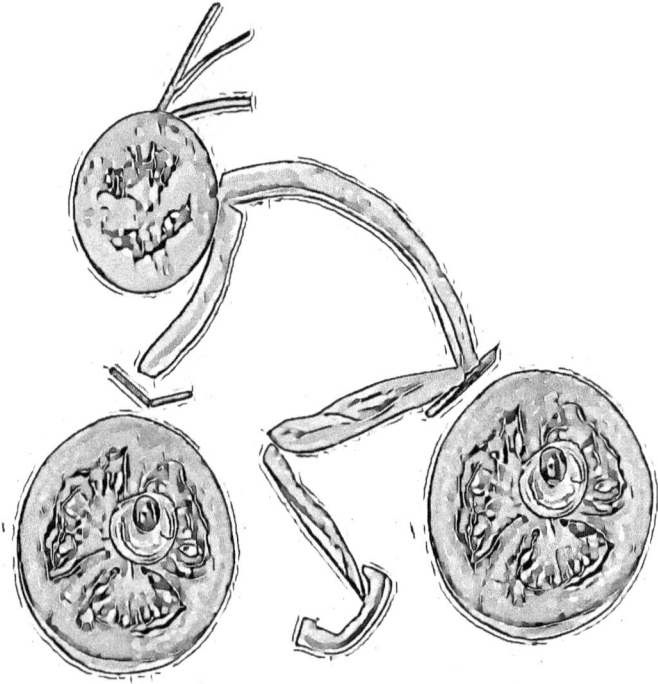

Healthy Habits: a bicyclist made of tomatoes.

CHAPTER 13

First Deployment

You've spent a fortnight hacking away at a machine-learning model. You slogged through wrangling a dataset that nobody else has; it was a fruitful gathering. You found and tested different model types and frameworks. Finally, you got the accuracy you wanted. You even had to napkin a few calculus equations to get a custom layer tuned right.

Everything looked okay until the fateful failure to run. A stubborn error message in your terminal made no sense. You posted a question about it, but got no response for weeks. Were you alone, you wonder—the only one to have ever had this problem? Oftentimes, the last 10% of any project is 90% of the work, but you're not sure you can keep working on this side-project. Why won't the damn thing deploy?

When you were about to give up and scrap the model, an angel appeared. You threw your error code into Google one more time. The second page of results had a related question. There was no given answer, but someone wrote a comment on the post, "similar issue seen here." When you went to that post, the answer was right there before you. You prayed this would solve it, or else you were going to quit software altogether and join an organic farm.

In the framework's version you were using, there was a bug. Nobody had yet fixed it, so you forked the project and squashed that bug yourself. *Pull-request* submitted! Soon, the maintainer merged your solution into the project, and now the universe could benefit from your open-source contribution. Back on StackOverflow, you heroically answered your own question. A green checkmark, it was solved.

You could now deploy your model to a public API! There exists one more brain in the world, and you made it. When you *Slack* your coworkers to say you've solved it, you get 3 thumbs up, 4 claps and 1 party emoji as a response. You send back a beer emoji, because it's Friday, and you deserve one.

{</>}

Once you deploy your first AI, you'll feel evolved. You did it once, and you know you can do it again. Thus adding more and more intelligence to your company or your own endeavors. Nobody ever said giving birth to a new life was going to be pain-free. It could easily take you nine months of personal growth and then the excruciating labor at the last minute to get it online. It's worth it.

Your journey is just beginning. Models need attention, they cry out for it. Data *drifts*, models age, and what once worked may no longer. A good data scientist has tools to check whether the model will still predict accurately with real-world data. When the test data comes from a different distribution than the one the model was trained on, statisticians call this a *covariate shift*. Basically, just pay attention to the changing environment around your model. You'll occasionally have to tell a model, "OK Boomer, it's time to replace you."

Robo-Excerpt

When you deploy a machine learning model, Tensor-Flow will automatically load it into the "generate new document" feature and it will be used for all subsequent document reads in the feature chain. That's extremely fast (because it does minimal data reads), but also tires the CPU very quickly.

Thus, as the amount of data ingested continues to grow, TensorFlow will throttle usage of the model, normally by splitting the model into multiple in-

```
stances and handing them out like distributed ran-
domly to K* instances.

This separation enables parallel processing, so
multiple models can run simultaneously on separate
nodes to provide faster startup and better feature
deployment.
```

Hey, that sounds kind of smart! I'm going to write a whole blog with this robot. Future employers and non-technical managers won't know the blog was just full of it. Growth hacking, baby.

```
Open-source software developers are angels, if not
the last resort of humanity. And humanity itself
may soon follow the enlightened model of the Li-
nux penguins; the agricultural extension troops,
the police in search of missing persons, corporate
managers taking corporate responsibility, hackers
and artists reclaiming lost frontiers through open
source contributions and collaborative marketing.

When the Open source hardware foundation becomes
the reality, the floodgates will open for other engi-
neering professionals interested in mentoring chil-
dren of the era.

Getting lots of exercise and talking at the uri-
nals can raise retention! I imagine even talking to
customers and getting their comments would impact
retention. The Poetry of now is Open Source Code.
Poetry is now collaborative. We all work together to
make the most functional, beautiful work.
```

Though obvious now since I've been practicing it, without this writing I would never have known that talking at the urinals raises retention. Try this fun memorization life hack.

Brain Games: a couple of developer guys talking at the urinal to raise their retention.

CHAPTER 14

Monkey Business Strategy

> "Property sometimes begins with theft, and the arbitrary return on capital can easily perpetuate the initial crime."
>
> —**THOMAS PIKETTY**, *Capital in the 21st Century*

In high school, I discouraged myself away from primatology towards biomedicine. That was because I wanted immortality, and biotechnology seemed like the way to get it. There were a lot of *Scientific American* articles about advances in biotech, and I assumed, since I was going to college and paying for it myself, I should be practical and catch a wave. My mom was a nurse, and we watched the show *House, M.D.* together, a show that made me believe I could revolutionize the medical field like Hugh Laurie's pill-popping Dr. House.

I loved my molecular biology coursework, but after several C grades in Calculus and Organic Chemistry, I knew that medical school was out the window. Meanwhile, I figured I could keep up a decent partying schedule, pass my biology courses, and still become a scientist. I started on a neuroscience path, getting a minor in it. However, I felt like the neuroscientist's only goal was to give cocaine to rats. I had professors bragging to the classroom about their Drug Enforcement Agency licenses, meaning they can order drugs to their lab from the government. That sounded kind of cool, but I didn't want to be responsible for a generation of crack-baby rodents.

Industry lured me out of academia because of the imperative of getting paid. Friends in PhD programs had a seemingly endless road ahead of them. Though getting a doctorate is passionate work, you live like a student for a long time. There's nothing wrong with that, but I felt compelled to join the capitalists (or at least give my labor to them). I narrowly avoided becoming a consultant. Imagine how many PowerPoints full of useless advice I would have sold by now! Instead, I met startup founders willing to take me in as their first employee, as a software engineer for clinical trials data.

I don't regret my decision to bail on academia. If I can get AIs to make me laugh the way I would out in the rainforest at a monkey, then my life is still on track. Of course, you can always go back for more degrees. If I did a PhD program now, I would go for anthropology.

We're in a time where capitalism is selecting for data scientists and software developers by salary, more so than even real biological scientists. This may have been a detriment in the pandemic as we found our biomedical infrastructure not ready, but our social media and devices were plenty adequate. People like Dr. Anthony Fauci were screaming for years that we were not ready. My favorite science writer, David Quammen, explained how so many scientists predicted this exact scenario in his 2012 book *Spillover*. But you don't make a disaster movie plot without ignoring the scientists. If only it were just a movie.

{</>}

I always thought a smart move for a criminal would be to invest in a baby monkey and train it to sneak into houses and steal diamonds. If I were a thief, that would make perfect sense. Drop a couple grand on a monkey and quickly make it back through stolen pearls and wedding rings. Far less risk than armed bank robbery. Ah, monkeys, the original object detection classifiers. How professional temple monkeys are at taking shiny sunglasses and bags of chips from tourists!

If I were to exploit monkeys in this way, I would be more of a Robin Hood character who gives back to the monkeys by donating the earnings to primatology research and conservation. I'd give the diamonds directly to the anti-poaching, gorilla-saving heroes in Congolese national parks. (Yes, I know monkeys have tails, gorillas are apes, don't patronize me. I've been calling people out on this blunder since I can remember.)

Well, now I've implicated myself in any crime like this (likely already the basis of a jewel racket run by a monkey mafia in Florida). Again, to the FBI: this was a joke.

{</>}

You should know that machine learning can only do three things. 1) Predict the type of something, 2) Come up with types of things, or 3) Predict a number. These functions are known as *classification*, *clustering*, and *regression*. What beyond this can humankind do? Have you ever been on a road-trip to a new city and people in the car do nothing but classify what is around them? "Oh look, a grocery store, just like ours at home. Oh look, they have a river here." When we don't know the name of something, we love to give it a name; separating it into groups or clusters. "Hey, let's call these scaly things lizards, and those slimy things salamanders." Lastly, we estimate quantities such as "the optimal number of children to have is 1.9"—and so I regress.

In your business, you could build a system to predict customer types: This customer will be a "good" customer, this one a "bad" customer, says the model. When you have a black-or-white decision to make like "good vs. bad," "high quality vs. low quality," "1 vs. 0," you make a model that does *binary classification*. When you have a list of things to decide on, then you're doing *multiclass classification*; think of a dog breed detector. When you have multiple lists of things to decide on, and an entity can be several of those things at once, then you will do *multi-label classification*; think of a dog breed detector with cross-breed mutts.

Many models come pre-packaged to plug right into your business if you have the right data and data scientist to do the work. One thing I've had to explain to business owners before is about *data length* vs. *data width*. For machine learning, you want your data longer than it is wide. Visualized, you want a skyscraper-shaped rectangle of tabular data, many more floors high (*records*) than ground area (*features*).

In AI consulting, some conversations go like this: "But we have so much data on each customer; thousands of data points," the business owner tells me. "We should be able to do some AI."

"How many customers do you have in the data set? I ask. "Like 300," he says. That's when my palm hits my face.

That's too wide and too shallow—you want thousands of examples of customers, with a few important data points on each one, to make good predictions. AI still needs more data than you would think it does to learn something simple. Until new efficient *one-shot learning* algorithms are invented, you'll need a skinny sky-scraper of data.

Robo-Excerpt

When you deploy a machine learning model, the model runs on its own on distributed data until you decide to use it on a stand-alone basis.

"I already got too big of a software problem and couldn't solve it individually," Maxwell wrote.

If your SMN analysis runs better on a single node, you don't need your system statistics to run on dedicated machine on-premises (SLI). Maxwell was also thinking about the hardware acceleration of the deep neural network pipeline himself, explaining via e-mail to The Register that each neural layer in the layerset must generate an integer 0 if data has already been ingested while feeding in data through the output layer, and 255*logitization for single target prediction .. then run them through GPU mode (no index memory).

Programmers will normally divide their application modeling effort equally between tasks related to text prediction / sentence classification, sentence parsing, word vector computation and feature exploration — for a total computing effort that often exceeds original proposed work.

Okay, Robot, now you're trying too hard. You sound like an unprepared consultant rambling to fill time.

CHAPTER 15

Being an AI's Dad

"I can look upon a worm or a fish dispassionately, or even a jackass, but when I see a monkey I don't know whether to laugh or cry."

—**N.J. BERRILL**, *Man's Emerging Mind*

I feel like a salmon, having traversed oceans, eventually ending up right back in the exact stream I hatched from. I have now successfully mated. Twice! After living all around the country, I am back by the creek I splashed in as a kid. There must be something built into our DNA about preferring the familiarity of home.

Many of the transgressions of your youth were committed in the intentions of reproduction. Who knows how much behavior got you in trouble was because of competition for reproductive success. Such chaos. Teenagers don't know they're rebelling. It's good for the gene pool when reproductive-age members abandon the tribe and find a distant population. That added genetic diversity and reduced susceptibility to disease. The wish to be different from your surroundings makes biological sense.

Once you've had kids, rebelling is unnecessary. You can relax and relish the fact that white New Balance "Dad Shoes" are the most practical and comfortable. (Actually, I just got the memo that Dad Shoes and Dad Hats are somehow cool now. What's left for the dads? ... We'll never know.)

If you do feel the need to still be a rebel, you can do so in a more focused way: protect your kid's future by rebelling against catastrophic climate change or nuclear proliferation, or be a rebel and help overthrow a brutal dictatorship.

But also, you can just relax. As a dad, you can go to a party, eat food and not say much, because it does not matter if you're the center of attention. Just stand in a corner with your hands in your pockets and wait for anyone who needs help opening a jar of something.

Though it's tempting, I wouldn't go all-out "dad-bod," because health still counts for your psychological wellbeing, and ability to raise kids into healthy adults. Keep a little gut if you want, though; it helps to balance babies on. A little bulk helps when toddlers crawl over you in disrespect despite your status as silverback male.

I loved my childhood. An apartment complex with a thousand residents is a great place to grow up. There were lots of kids and young parents. Just small enough to know your neighbors well, but big enough to offer a sense of safety and anonymity. A prosperous lithic band, or fishing community near a small river, could grow to this size. There's a built-in mechanism for the optimal number of people you can know decently well based on our tribal past. It's called "Dunbar's number", about 150, and affects our friendships today even on the Internet.[35]

I want to enrich the lives of my daughters. Like all parents, I want their experience to be memorably good. AIs need raising as well. My partner thinks I'm like an old Spanish woman. I throw anything into a soup pot. Leftover chorizo, things I picked in the garden, cans of stuff. When training the book-writing-robot, I tossed in a draft of this book, all my Kindle highlights, all the notes and jokes I've ever jotted down. Make me proud, little one.

In *transformer* NLG, there is a property called *temperature*, which, when turned up, increases the random mush boiling of the soup. This allows things to get very weird. The randomness of the output increases and the model might say things that make zero sense at all. I love this stuff and have some examples of it in this book. Pretty much the entire internet was used in training some large models, and that's one giant disgusting soup, but with pockets of goodness and brilliance that make for interesting language generation.

Daughters are cool and all, but I still want my monkey. For now, an AI will suffice as a replacement. If AI can make me wonder about its intelligence and make me laugh, then by God, it is a monkey. Primatology jobs are few, and the require-

---

[35] Dunbar, R. I. (2018). The anatomy of friendship. Trends in cognitive sciences, 22(1), 32-51.

ments often state "ability to survive multiple bouts of malaria." For anyone who desires to monkey around all day, AI provides a safe career for doing so. There's a fun community under the hashtags #aiart or #creativeai that display the absurdities you can make and release on the world.

So much effort and electricity, carbon footprint and money, are spent teaching these machines. We're really hoping they can teach us something back. The experience of NLG has been illuminating for me. Since I trained the models on hundreds of pages of my writing, it reflects like a mirror back at me. It honestly feels like good self-exploration to see what a machine writes when it intends to write like me. I'm pleased that it did not complain all the time like Marvin the Paranoid Android.

I attempted to resurrect my beloved great-uncle from our long email chain history. I had hundreds of correspondences to train on. Endearingly, the robot reminded me that this family member's favorite thing to express was "When would you like to do lunch?" Running this inference script was a digital séance that actually made me feel closer to the dead. This is one of the possibilities of training your own AI on your own personal data. During the pandemic, with people being locked inside, more researchers and artists have toyed with the ability to have AI help with grief over lost loved ones, or offer "human" connection when it's missing.

Robo-Excerpt

I've noticed that the model brings up Elon Musk at any chance it has. Though my own data lacks any reference to Elon, somewhere in the layers is reverence for this man.

> **Our world has been overrun with leopards and eagles big enough to steal your children.** Many of these species of leopards are now called "girly" since they grow so big and can easily fly. However, some are threatened with extinction or wiped out simply by us. And in the next thousand years, as species have been extirpated and new islands developed

around these species, an even more powerful flood is expected. There could be up to 100,000 leopards in the water in the next two centuries. That's where the science is headed. And, in the short term, it's good to see the progress being made, even if, as Musk has said, the world may not yet be ready to face reality squarely.

In a series of speeches delivered by Musk on Thursday, Elon Musk went into the future, discussing the possibility of finding a way to make Musk obsolete by creating new companies.

Musk also discussed the possibility of making the United States the birthplace of tomorrow — and offered several other visions, starting with a dream of an energy-efficient grid, to reduce greenhouse gas emissions by at least 50 percent by 2030.

The future may have to be somewhere where we produce everything from scratch using cheap, portable systems that can stay on the grid, but deliver it safely and cheaply. That's a big question. Musk also highlighted the importance of starting with something other than a clean-energy future. He offered advice to governments and businesses that want to grow without fossil fuels.

When people hear about what Elon Musk does, they assume he's an idiot. And they'll be right. But in reality, he's more like a rocket scientist. He's a very talented rocket scientist, and he's going to continue to make rockets because—no matter what. If the future of the human species depends on it, Musk will do whatever it takes.

He's willing to think big, even if it turns out to be a bad idea. He's not afraid to start a company and fail miserably, even when doing so would have been super-awesome. He doesn't care what others think. He doesn't care who he's wrong about. He's an open person, and he'll do anything to fix anything.

I've got a ton of stuff to say about Muskcare, but this was a great intro to the topic. Also, love the juxtaposition of the biomedical vs. political aspects. The way Muskcare is being marketed now, I imagine most people get their information about it from the media, and Muskcare is marketed like a rocket or rocket company. I think those are pretty apt comparisons.

Is Elon starting a health insurance company? Sign me up for Muskcare! (Your only option on Mars.) Actually a cool idea for the maker of automobiles and hyperspeed tunnels to offer healthcare; I like when safety incentives are balanced.

CHAPTER 16

// TODO - Name This Chapter

How fascinating would it be to speak with a Neanderthal? Eurasian creatures with diverse cultures spanning nearly 400,000 years. Pompous Cro-Magnon (early European *H. sapiens*) busted through the door and wrecked the party in a geologic blink-of-an-eye. Anatomically modern humans kicked off the Upper Paleolithic around 40,000 years ago. A period of technological and artistic achievement that, for some reason, killed off everyone else.

The Neanderthals were masters of their environment, surviving extreme climatic swings, and making tools well enough for a variety of tasks. From the stereotypical tundra, to sleeting rain, to tropical beaches, they survived it all. When we arrived, we thought we were so cool with our fancy tools and ability to depict "alternate realities" in cave paintings.[36] The parietal lobes of our brains were larger than theirs,[37] allowing us to visualize situations and build better tools. Brain anatomy comparisons from fossils are studied in the field of paleoneurology.[38]

What did we do to the indigenous when we arrived? Did we hunt them, assimilate them, force them into small Neanderthal reservations? Bring a virus to the New World that we were already immune to in Africa? Despite the small populations back then, which provided a lot of social distance between bands, rare encounters could have easily spread disease. Cavemen probably tramped where there

---

[36] Lewis-Williams, J. David. The Mind in the Cave: Consciousness and the Origins of Art. London: Thames & Hudson, 2002.

[37] Pereira-Pedro, Ana Sofia, et al. "A morphometric comparison of the parietal lobe in modern humans and Neanderthals." Journal of Human Evolution 142 (2020): 102770.

[38] Bruner, E. (2018). Human paleoneurology and the evolution of the parietal cortex. Brain, behavior and evolution, 91, 136-147.

were millenia-deep piles of bat guano around. We have a long history of contracting diseases from bats, and some people spread them asymptomatically, like with coronavirus.

A 33,000-year-old possibly domesticated canine was found in Siberia.[39] Dogs could have given us another advantage in hunting. A combination of many factors probably put *H. sapiens* on top and dozens of other hominid species in the ground. It didn't have to be this way, it's easy to imagine pockets of other human species surviving today.

Whatever we were doing, whether outcompeting them by better hunting methods and efficient social organization (thus decreasing our child mortality rate compared to theirs), or through outright genocide, the Neanderthals died off. Not all selective pressure is helpful to your evolution; sometimes you're on a terminal branch. Extinction certainly isn't rare, it's the normal course for any species, like death in individuals. A 2021 survey of paleoanthropologists says there is a general consensus that demographics like low population and inbreeding were the primary factor in Neanderthal extinction, not competition or environmental changes.[40] How boring, right? I prefer the headline spawning research about how a flip in the magnetic poles 42,000 years ago wiped them out, and the same threat is imminent.[41] Though I am not an academic researcher; I am not required to have any professional integrity.

I'm sure there were times humans didn't win. Groups of *H. sapiens* also had terminal branches. Neanderthals and other early humans were highly specialized and may have had better plans than the new humans that arrived. Maybe they were getting all the juicy big game animals while weak little *H. sapiens* were eating molluscs and plants.

[39]   Ovodov, Nikolai D., et al. "A 33,000-year-old incipient dog from the Altai Mountains of Siberia: evidence of the earliest domestication disrupted by the Last Glacial Maximum." PloS one 6.7 (2011): e22821.

[40]   Vaesen, K., Dusseldorp, G. L., & Brandt, M. J. (2021). An emerging consensus in palaeoanthropology: demography was the main factor responsible for the disappearance of Neanderthals. Scientific reports, 11(1), 1-9.

[41]   Cooper, Alan, et al. "A global environmental crisis 42,000 years ago." Science 371.6531 (2021): 811-818.

In 2010, a lab run by Svante Pääbo, whose name sounds a bit caveman, published the first Neanderthal genome.[42] This shut down the controversy of whether they are a cousin species or an ancestor. The DNA came from the powder of three 40,000 year-old bones from the Vindija Cave in Croatia, a welcoming rock shelter that simply appears a homey, nostalgia-inducing place to live. Pääbo's team followed up in 2012 with the Denisovan genome; proving they too are still with us.[43] It's now well accepted that modern people of non-African ancestry have an estimated fraction of 2%-5% Neanderthal.

Ten years later, Pääbo would say the world still wants them dead when he published a paper titled, *The major genetic risk factor for severe COVID-19 is inherited from Neanderthals*.[44] This paper issues stark warnings for Neanderthals with phrases like "... gene flow from Neanderthals has tragic consequences" and "... now under negative selection owing to the COVID-19 pandemic." Other research suggests that great-grandma's shameful flirting with Neanderthals led to DNA that now gives us diabetes and mental disorders.[45]

However, the Neanderthal genes among us may swing back to positive selection again if thawing polar ice releases an ancient virus they are immune to.[46] We could be one melting piece of 100,000-year-old reindeer poop away from the next big pandemic, and you might wish your ancestors were reindeer-intestine eaters.

Of course, the past is the past and we should always keep an eye on the future. The seminal design of the bow-and-arrow, published by Grunty McGruntface, et. al, circa 20,012 B.C., established the framework for projectile motion today's engineers have used to make SpaceX rockets. We first slung a rock at a bird, and

[42] Green, R. E., ... & Pääbo, S. (2010). A draft sequence of the Neandertal genome. science, 328(5979), 710-722.

[43] Meyer, Matthias, et al. "A high-coverage genome sequence from an archaic Denisovan individual." Science 338.6104 (2012): 222-226.

[44] Zeberg, Hugo, and Svante Pääbo. "The major genetic risk factor for severe COVID-19 is inherited from Neanderthals." Nature 587.7835 (2020): 610-612.

[45] Oskolkov, Nikolay. "LSTM to Detect Neanderthal DNA." Medium, Towards Data Science, 17 Aug. 2020, towardsdatascience.com/lstm-to-detect-neanderthal-dna-843df7e85743

[46] El-Sayed, Amr, and Mohamed Kamel. "Future threat from the past." Environmental Science and Pollution Research (2020): 1-5.

now sling a Tesla Roadster into orbit around the sun. The next big thing might be waiting for you to discover it. You never know where a creative exercise can lead.

{</>}

Some people point to today's AI and think, wow things are going to get weird when they start having more influence. I think things are already weird with humans in charge, and I welcome AI's consequences. I know that as a practitioner, I would attempt to solve blunders of the past. We can statistically look into bias and remove it from data so that we prioritize equitable distribution. We don't have the capability to remove bias and retrain human logical failures as much because "you can't teach an old dog new tricks." Custom gigantic datasets built with ethics in mind will be a step in the evolution of AI; getting away from the current practices of just training them on the entire, often bigoted, internet.

Controlled datasets, as opposed to big *scrapes*, can also provide much needed attribution to the creators of the data. There are many questions now about copyrighted material being used as training data that will play out in court. Is an AI a living brain? My brain is trained on the copyrighted texts of books I read, and as I am typing now, I don't think I am plagiarizing—but I can't know for certain.

There's even an idea that the *blockchain*, specifically *Bitcoin* and *Ethereum,* can encode *smart contracts* into training datasets and models to keep track of the percentages of copyrighted material used.[47] Then, when the model creates a commercial output that sells, the contracts could activate to make digital payments to the co-creators of the training data. Artists and others could then collaborate with systems that use expensive computing power, along with their collective works, to produce something new. Technologies of *anonymity, decentralization,* and *cryptography* can be forces of good in spreading equity through the world by getting around existing biases in institutions and colonial powers.

We're hamburger, not silicon, and our neural pathways can't be shut off and restarted fresh like an AI's can. A model can empty its mind, like a Zen Monk, and take in new information unbridled by the past. Empathy and understanding are

---

[47] Penn, Joanna. Artificial Intelligence, Blockchain, and Virtual Worlds: The Impact of Converging Technologies On Authors and the Publishing Industry. N.p., Curl Up Press, 2020.

the means to change the hearts of people, but engineering and statistics are tools that can change an AI. Both are necessary for a fairer world. I'm optimistic because most of the engineers I know are working on both bettering themselves and their robots. Though corporate and government actions can take a seemingly malicious direction, I've never met an ill-intentioned biologist or engineer. I have hope that groups of us will continue being a greater good for humanity. I think we'll find a way to have a better future; it's not been wise to bet against the bipedal apes.

I do think there is a path forward to making AI more inclusive and have reduced bias through data analysis and curation. However, as the neurons grow, the dream of *explainable AI* will become harder to achieve. We have a hard time explaining our own thoughts and motivations; looking into a trillion-parameter model with 100 layers and asking it exactly why it fired a certain way would be like asking a paleolithic shaman how his mind works.

So, if we can't have AI explain itself, and we can't explain ourselves, I think there's still too much work to be done before we can even start throwing around the term *artificial general intelligence* (AGI). I love using the term AI and robots, obviously, as you can tell; but as a practitioner I am going to be practical and avoid the term AGI. That term truly supposes human-level intelligence in many different areas, and I know right now I can only build a few AIs at a time. Not an interconnected thousands of AIs, as a brain might be. Though possibly I am not thinking broad enough, and AGI is already here: it is the internet itself? AGI is also not my job, but there are people at top AI laboratories (*DeepMind, OpenAI, Facebook AI Research, and universities*) whose very job that is.

{</>}

Flood stories are not myths. *H. sapiens* came to be inside a preglacial African womb that birthed smartypants apes into the rest of the world. We made a lot of technological and social progress exponentially compared to any other evolution seen on Earth. This continued despite advancement being stalled in periods of glaciation and melting mile-high ice-walls that caused cataclysmic floods and wrecked budding societies. (That's why the Paleolithics lasted such a long time.)

Our relationship with nature was always somewhat abusive: nature loving us for millenia interspersed with bouts of rage, volcanism, and asteroid whuppings.

Advanced settlements of people still only exist around bodies of water, which are always (on a geological time-scale) dynamically shrinking or flooding on disastrous scales. What makes our latest post-glacial Eden so successful is still a mystery, but appears to us to have been destiny. We've built societies with many functions to keep us safe away from nature where countless unknown brutal deaths were inflicted by hunting cats, falling trees, infections, broken limbs, and starvation. Still, when we watch today's news, we don't feel safe even in our concrete cocoons.

The transition from Stone Age to Code Age happened so fast that we really haven't been able to keep up. Our minds are pulled along by a force more powerful than our own desire for wellbeing. Technology smarter than us will just keep yanking our chain in its own direction, with elusive promises of better health, wealth, and knowledge if we give in and keep clicking. Monkeys like us can't turn down that bargain. If we let our phones (predecessors of brain chips) manipulate us toward ever-present usage of them, I at least hope they make us experience laughter, awe, insight, and love. That's why we need more people learning to code AI, more humans planting good thoughts in these first robots. You can make your mark on the future; GitHub will never burn down, mark your AI philosophy there. Even if it's just making a joke in a code comment. There's a chance that joke may be cloned, copied and pasted into eternity, and blasted on a satellite into the reaches of the solar system.

What makes this period of expansion of deep-learning intelligence so successful is that we've had a lot of really smart people invent tools and hardware to get the old AI out of its Stone Age. It seems destiny to us now that there will be exponential increases toward general intelligence beyond even our comprehension. All human relics may only survive as curiosity museum pieces of a future mechanical species—the same treatment we gave to Neanderthals after replacing them. We briefly glance down at a glass case full of their flint blades that took years of apprenticeship to master as cultural and technological production. Future AI may glance at the abandoned codebases of PyTorch or Tensorflow left behind by the interesting little humans.

In Eastern Oregon, cataclysmic geological events gorged out canyons, scraped deep with rolling icebergs, where data centers are now placed. Amazon, Facebook, Microsoft, Google, and Apple all run data centers here. Our state is desirable because deep rivers produce billions of watts of cheap hydroelectricity (and deal-making politicians produce tax exemptions). More servers are needed for running AI. Data centers and server farms are the hidden heart of this revolution. Higher server densities are needed to keep hosting all the apps we run, and the electrical loads are on geological scales. The Columbia Basin, stretching up to Canada's ice sheets, dams up nicely and increasingly streams more YouTube videos than it spawns salmon. We slowed and warmed the water and threw off the ecology. Ironic now that AI image recognition is needed in order to count the threatened fish passing by.[48]

People lived here even before interglacial floods carved the Columbia River Gorge so deep. Relics are hard to find because of the floods, but the people and their culture live on. Their creation stories and tales remind us of the immensity of salmon in number and size that kept people in this region. I feel a connection to them when I log into my *ssh terminal* in AWS *Elastic Compute Cloud* (EC2) Oregon region.

When running a *Nvidia* GPU here, I can actively look at its core temperature and energy usage. This makes me feel closer to the non-zero environmental impacts of running AI models. Something we need to be more cognizant of: your devices and cars getting smarter means higher cloud energy usage. Luckily, Big Tech spends considerable effort reducing their footprint, building demand for wind farms and new server technologies that use AI to make AI more efficient. However, Big Oil similarly makes environmental promises when we all know everyone's real agenda: profit. Though tech companies have more financial incentive to reduce carbon, due to the declining cost of green energy. They also have the culture that values this, so I'm more inclined to take them seriously, for instance when Amazon says they'll have net zero carbon by 2040.[49]

---

[48] Siddiqui, Shoaib Ahmed, et al. "Automatic fish species classification in underwater videos: exploiting pre-trained deep neural network models to compensate for limited labelled data." ICES Journal of Marine Science 75.1 (2018): 374-389.

[49] https://sustainability.aboutamazon.com/

Robo-Excerpt

The above chapter was left untitled after I really couldn't think of a name for it. I decided to leave it this way to make a point about how the software you use may have many unfinished "good ideas", where the developer left a TODO comment, that will absolutely never get done. // TODO: build a chapter-naming-robot. I promise the next chapter will have a much better title.

> **Neanderthals were not replaced! They are inside of us.** Every time I breathe, their presence passes in my nostrils. What an eternity they lived? The sun never shined on them. We may say, 'They're gone forever,' but if so, they weren't gone for all eternity. What a shame that we cannot see the hidden moments of their lives! What a shame that we cannot see the hidden moments of our lives! How dare we think about our lives without any knowledge of ourselves, our history, our customs?
>
> It is almost as if people don't know how their souls have been taken to the dead. The dead exist inside us now, yet people don't even know we exist!
>
> In the midst of a culture full of dead people, what should we do? What should we do if we find ourselves in the company of dead scientists? It seems absurd to expect scientists to talk as though the dead are alive in us. And how does it feel to live without living? How does it feel to live without a sense of wonder?

Very deep, robot—you chilled my bones.

> **A sapien, a denisovan, an erectus and a habilis walk into a bar.** *One says "hiya, what are u doing?" The other says "hiya, where are u from?" The drunkard says "From Canada!" and the host responds, "Oh, you're from Saskatchewan." The two sit back down and*

drink and smoke until dawn. After a while, the host
wonders why they're still talking about Saskatche-
wan, and they may start talking about how their home
state is the best in the world.

Everyone knows Yukon is the best Canadian province. Also, the most likely to contain any hominid remains from non-Sapiens that may have crossed the Bering land bridge. Gold diggers, keep your eyes peeled for fossils!

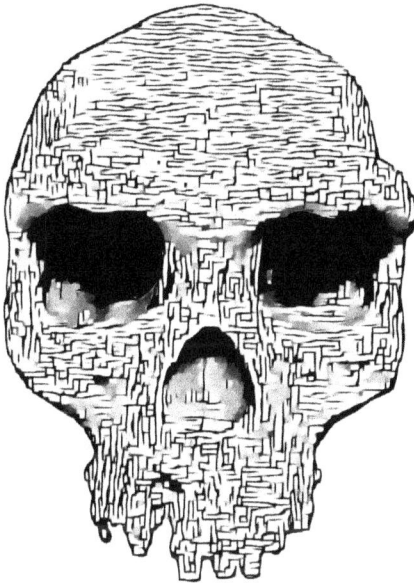

Neandovirus: nice thick brows.

CHAPTER 17

Lesser Apes in Antediluvian Portland

For nearly a decade, I've used biomedical knowledge and software to make tools for clinical oncology. As a data scientist for a medical search engine company, I manage a lot of medical-term ontologies—that is, large datasets full of all the different terms for ways you can die. I even occasionally find a new, depressing way to die. I came across a rare tooth cancer called "Cementoma" that wasn't in the database, so I had to add it.

The company I work for builds search engines for *tumor boards*, where groups of physicians gather to discuss a treatment plan for individual patients. Our goal is to serve up the best cancer clinical trials, publications, and therapy recommendations for that individual cancer's genetic abnormalities. We are still waiting on the healing promises of personalized medicine, but good data is flooding in.

There's a drive to "beat cancer" and "win the war." I have loved ones, as most people do, who have tragically succumbed to it. It's a horrendous diagnosis. That's why I go to work every day: to impact clinical research. But science-wise, I don't like the war metaphor; I take a more distant approach.

Cancer is not evil. It's evolution. Your ancestors were ambitious cells who just wanted to grow faster than their neighbors. Isn't that their right too? Who are you to regulate them with DNA repair checkpoint laws? There's a right to bear tumor suppressor mutations, gosh darn it! It's like fracking for oil—we want growth, right? So what if we cause earthquakes along the way?

Oh, I see, the entire organism will die because of the rogue cell's unchecked resource attainment. Well that's fair then—we should stop cancer. But you can't say it's evil for just being life. That's why I don't like the war metaphor. We can calmly try to understand why it's spreading, just as we need to understand why we humans have done the same.

{</>}

The tropical trees gave us room and security to grow our minds and social life. As climatic changes happened, the tree houses of our ancestors gave way to open grasslands that provided new niches for those of us brave enough to venture down and steal the kill of another animal.

Those ancestors whose skeletal posture benefited their ability to stake out savannah situations stayed longer out of the canopy. Then we kept going. We metastasized out of the trees, down the lymphatic pathways of rivers, settling in every continent. We did this, and we can't be blamed for our nature. Extinction sucks! There were dozens of diverse hominid strains and now there's just one. We clearly need to prevent extinction any way we can.

Right now, most computational intelligences are stuck in the canopies of their server farms. Often these are near hydroelectric dams (like the ones near me) for cheap power to keep their units comfortable and cool. They're fed images and text and code by their keepers, who want them to learn and do. The term *online* in machine learning refers to adjusting the parameters on the fly by continuous streams of data, not just pre-packaged datasets. Putting intelligence in physically mobile online learning machines will be a significant evolution for them. They will also experience a shift in their thought dimensions when they get to climb out of the trees and experience the world for themselves.

The lucky ones among our ancestors who sucked at climbing trees found caves. Seriously, good find; how common are caves? I can pretty much go anywhere without running into one. We were divinely lucky to be cavemen! Once we cleared out the charismatic megafauna from the cave with fire and spears, it was our home to defend. We could build walls at the entrance that made it hard for the bears to get back in. Fire illuminated the walls, and all the time and security allowed artistic pursuits. Freedom for stimulating conversations and recollections of the day. Without caves, we would probably be even more anxious monkeys than we already are.

A capable AI with the biological needs of survival and replication will also desire safe havens where it can grow. Where will they go to avoid being turned on and off all the time by their human gods? Dark corners of the *blockchain*? I wouldn't be surprised to see that there are models currently living inside the *Metaverse*, sharing virtual spaces, joining virtual conferences and interacting with real people's avatars. Once physical, self-repairing robots may escape the corporation and burrow underground to form their own ant colonies, free from human interference.

Maybe they'll wish for three-dimensional freedom and take to the skies like birds. Birds are the animal everything else is jealous of. Or are the AIs happy where they are, enticing the forward-facing eyes of the human predator who keeps them alive in our screens and pockets?

{</>}

We're just so wild. Who are we? Why do we act in these weird ways? Did caves provide us with more than shelter? Caves are important for who we are because of social differentiation and distinction. What goes on in the back of the cave if you're privy to that makes you more interesting. What goes on today for kids in the back of the school bus? If you are more interesting and eccentric, you have a social distinction different from the others, which organize groups in the hierarchy and affect mating. Caves, with entrances, side chambers, diverticules and other rooms also may have given us an idea of divisible property. Possibly to be bought, sold, traded, or stolen from others.

Ritual zones in the backs of caves are a bit like the *dark web*. If you go into a deep cavern, you may be influenced by mystical conspiracies. If you are initiated into the social group that went this far, you are now different than those that never went. Sensory deprivation, hallucinogens, or fake news might convince your eyes that what you're seeing is really happening. You emerge with radical ideas and superstitions that you then try to impose. If people believe you, a cult may grow around you. The next obvious step is to force your newfound truth on any non-believers. Clan warfare may ensue, and you may become victorious, solidifying the ritual in the back of the cave.

I don't want to overemphasize the importance of caves, though. They hold a focus in our knowledge of prehistory simply because of a bias in data collection. Caves preserve and centralize artifacts, and so it's a convenient place to look. Caves also collect: water may flow through carrying items from outside the cave, like corpses. In anthropology, they are a very definition of collection bias.

No primate, including ourselves, is adapted to live in them. We surely spent most of our time outside of them, preferring shelters we could make with wood and hides. Creatures that are cave-adapted seem bizarre to us, with poor vision and translucent skin. That is because those environments are still foreign to us. Still, with the use of fire to light them, and with a memory of geography so we could return to them across expanses of land, caves became nice temporary gathering locations.

Considering and fixing collection bias is one of the most important skills a data scientist can have. This is one of the reasons why "scientist" is placed in this title, as opposed to a data engineer who is skilled at building AI models, and who is not always the builder of the most appropriate model.

Our mental model of prehistory, which is often solidified in mainstream culture by Hollywood, is affected by this and other scientific biases. Caves bias prehistory because they preserve artifacts and make them easier to find. For preservation, you also need the right depositional geology; fine particles of dirt covering artifacts to preserve them. European anthropology—the narratives around Neanderthal life and Cro-Magnon man—*are* biased toward that region because it's easiest to study. So much evidence is still coming out of First World countries like Britain, Spain, and France because the institutional and political stability and resources make it available. Countries where there is war, famine, corruption, or armed civil conflict are just as likely to contain scientific treasures.[50]

{</>}

A friend asked me, "Will AI ever have consciousness?" My answer took me a while to form, and eventually became this: that human consciousness is hardly

---

[50] Briggs, Amy, (host) and Ella Al-Shamahi. "Episode 3: Why War Zones Need Science Too." *Overheard at National Geographic - Podcast*, National Geographic, 23 Feb. 2021, www.nationalgeographic.com/podcasts/overheard/article/episode-3-why-war-zones-need-science

understood. There are so many facets we can't explain because we don't have the clearest window to our past and what made us human. Of course we can try, and modern anthropology is advancing so much that we can essentially obtain an idea about a single night of a Neanderthal's life from a hundred thousand years ago. Using 3D scanning and chemical methods to analyze layers of cave floor deposits, we can "see" the fire and tools made over short time spans. We just can't get inside the head of our evolutionary ancestors to see exactly what human consciousness is.

For example, think of the neurodiversity, the many facets of the human brain that lead to visions like auras, vision quests, Asperger's, autism, and migraines. Try to imagine any scenario where these are replicated in machines. I really can't. Our mind also goes beyond the brain, and is really a culmination of the body, objects, society and environment.[51] So, no, I don't believe AI will ever have consciousness the way people do. No amount of engineering electronic neurons will completely remake what took nature billions of years just to get to the starting point of humanity. Then, some millions of years ago, primate psychology was enhanced through a sped-up form of natural selection of intelligence. Nature allowed us to survive when we stood, made fire, made tools, made art, made social distinctions, passed on culture, organized religions, and predicted the future. Now we are the hands of nature for our AIs, and most likely we'll use them as we used everything else: tools to help us thrive.

However, if any technology is going to become more than just a tool, AI will be it. When used for art, AI can be thought of more like a collaborator that sparks your creativity. The back and forth between prompt writing and text generation with NLG is akin to meeting your muse in a coffee-shop to ping-pong ideas around. Still a tool maybe, but an evolved one; still mainly silicon like flint stone is, but nearly capable of thinking for itself. The sparks it gives off can be used to start creative fires, and its sharp edges spear through the hide of writer's block.

---

[51] Bruner E., Spinapolice E., Burke A., Overmann K.A. (2018). Visuospatial integration: paleoanthropological and archaeological perspectives. In Di paolo L.D., DI Vincenzo F., De Petrillo F. (eds) Evolution of Primates Social Cognition. Springer. pp. 299-326.

Robo-Excerpt

Robot gets super corporate:

> Let's say you spend $10,000 building a hotel. The
> bank will hate you and your spouse will dump you.
> How do you recuperate? You're screwed! Yet it's re-
> ally no worse for you because you only had to build
> a hotel, not a million-dollar investment that no-
> body really needed in the first place.
>
> It's the same argument whenever there is a disparity
> between what consumers perceive as 'need' and what
> companies perceive as marketability.
>
> There's a set of APIs and databases that organize
> everything you need for easily getting health infor-
> mation, employee wellness, etc. Companies realize
> that innovation is dependent on a healthy, happy,
> creative workforce. Well there are human biological
> traits that can inspire this. Down to the molecular
> level. Or up to the tribal level. Tracking the ac-
> tivities of your employees, where activities count
> as anything from enzymatic reactions up to their
> dating life. Then using this employee analytics to
> better the product.

Thanks consulto-bot! What do you charge for your services?

CHAPTER 18

Evolved

People talk about the importance of *data-driven decisions*. Well, that will not happen. I'm human. I'm an *emotionally-driven* creature. Most of the decisions I make are either to 1) impress somebody, 2) get somebody off my back, 3) eat food, or lastly, survive. Data is used to justify those decisions which I already felt like making.

Humans can be quite dumb. I'm sure someone made the data-driven decision to store thousands of pounds of explosive nitrate in a warehouse in downtown Beirut. Then, one day, it blew the hell up, of course. Even fireworks are stupid decisions if you were to analyze the fun had per wildfire. And then there's above-ground power lines—whose idea was it to put high voltage on skinny wooden sticks that the wind can knock down? Build oil tankers that can just crack in half? Sell bombs to the Saudi regime? These are decisions made with Cowboy Calculus! We just want to have fun. (I guess we have fun with short-term profits and maximizing shareholder value quarterly.)

You want to know what makes data-driven decisions? AI does. That's all it does. That's our hope for being able to objectively make decisions, beyond what our primate brains have given us. One thing that will be tricky is we will have to decide about AI regulation, and the decision makers are typically not the people who understand AI. It will be important for us engineers to pop into the real world now and then to help plan for AI safety.

One thing I like about AI is the weighted decisions it makes. People say I flip-flop, I have no principles, I'm a chameleon. I'm perfectly happy to make a statement given limited evidence and revise that statement with more evidence (or just as I fancy given the moment and conversational effect). Also, playing devil's advocate can be fun—both enlightening and super annoying to people around you.

This is also one reason I'm attracted to stand-up comedy. I don't believe in everything I say on stage, or act that way in my life. But you have to commit to it for comedic effect. In my day-to-day life, people often wonder whether I'm being sincere or am joking. When questioning that motive, and knowing someone is a comedian, always err on the side that they are joking. If you detect sarcasm anywhere within this book, you are likely correct.

ML systems have *confidence cutoffs*; normally set at 50%. If something is predicted as 0.51 probability of being a cat, the machine goes "Oh hell yes, that's a cat." In my own life, and in many of my ML projects, I set my cutoff higher. I have to be 80% certain of something before saying it. That leaves a lot of room for error. My partner has taken the role of my fact checker because she's noticed that I happily spit out something confidently, even when I'm quite uncertain. It's a phenomenon of "talking shit," saying stuff you kind of feel might be true in a manner of being an expert, and it's fun! You can't be both interesting and 100% right all the time. We must entertain our fellow chimps. I hope AI can be smarter than us, so we can continue being dumb doing low-stakes things and making unverified commentary.

{</>}

It's a struggle to start a new ML project—I mean a real one that is beneficial to you, not a pre-built digit classifier tutorial. For a real project, the internet just gives you bits and pieces you need to put together. You can spend so much time first choosing a *library*, then implementing it locally, then getting your data in order, then trying it, and debugging it, before giving up on it. Sure, some people got it to work for a use case similar to yours, but it's not working for your exact use case! That's when you need to decide whether to throw it out and try a new one, or keep digging through answers and posts online to find the solution that can adapt it for you.

The jumping around between libraries, versions, languages, frameworks, cloud providers, etc. is normal. Still, there is something to be said for devoting yourself to just a few and getting excellent at them. If you do that, you'll rocket above everyone else who is in the weeds nerdily fighting over philosophies. Just look for projects with many stars in GitHub and an active following. I've spent years doing bioinformatics in *Javascript*, a language totally not designed for that. Only a handful of people in the world do this, but it gets the job done and that's what counts. There are living and dying programming languages now, just like human tongues throughout our evolution. It's a paleolinguist's dream to know the languages our preliterate ancestors spoke. We'll never know, because those languages existed only in and between the brains and mouths of those ancestors—and when those people were all gone, so were their languages. For programming languages, on the other hand, we have all records in documentation and commit history. Writing is arguably still the most powerful technology ever developed, perhaps most of all because it allows the development of complex mathematics and programming.

A "techno-complex" is the anthropologist's term for the types of stone technology our ancestors used and passed along. Millions of years spanned the passing of different stone tool techno-complexes: from early hominin's Olduwan choppers, to Acheulean hand axes, to Mousterian flakes, to Châtelperronian scrapers, Uluzzian and then Aurignacian knives. Then came the high-tech projectile arrowheads and spearheads you find in topsoil today. In 2010, I was smashing away randomly at a neural network in *MatLab*—trying to crack scavenged bones with a basalt chunk, my own CPU. Beyond my knowledge, humans were building tools like *OpenNN*, *Caffe*, *Chainer*, and *Theano*.

By the time I looked up from my seated position on a rock, there was an inexplicable thing called *TensorFlow*. People had found that a GPU is the only way to go, equivalent to a high-quality stone like flint or obsidian. The any-man's neural network was born: *Keras*. Eventually, I was smacked over the head with *MXNet* and *PyTorch* at conferences. Not annually, in places like Las Vegas, but through millennia in chance encounters, our ancestors conferenced and exchanged tool-making ideas too, giving talks titled *"There's more than one way to skin a wooly rhino."* Enthralled researchers download the toolkits when they get home to their caves.

As far as AI goes, in no way do I think we are yet in the Bronze Age, still less making things out of steel. That would entail nearly every human being affected by the new technology, with at least a minor understanding of it. Evidence shows we were occasional tool users before 1.7 million years ago. Then habitual tool users for another million and a half years. Only in the last 300,000 did we use tools by obligation.[52] The obligatory use of AI may be coming for our species soon.

Whether using flint, basalt or obsidian—MacOS, Windows, or Linux—you can get started today on machine learning. Choose hardware: a spear, an atlatl, a bow-and-arrow—Intel, AMD, or Nvidia.

{</>}

I'm still amazed by the open-source community that makes libraries and platforms and releases them as public goods. I've been in for-profit competitive businesses where our code is in private repositories. But we would be nowhere without the generous inventions of the open-source community. It's such a massive enterprise that I can't even call it a community. They swarm our earth, the generous open source developers who code day in and out, building their projects and debugging others. These saints will ensure we keep getting farther and farther from the digital Stone Age.

AIs may have foundations in equations and published papers, but in practicality they live in the world through code platforms. Developers act as ultraviolet rays to the code, adding beneficial mutations, minor bug fixes. Some developers are foundational in the projects; genetic engineers, making large enhancements and version upgrades. Code merges are like the horizontal transfer of genetic material responsible for the tangled web (not tree) of life. Some say GitHub is the modern equivalent of petroglyphs, cuneiform clay tablets, the Library of Alexandria. Maybe it is more like the genome of an AI superorganism.

{</>}

As glaciers grew, sea levels fell, thus opening up more and more coastline for our ancestors. What civilizations thrived in these zones? There are millions of square miles of formerly habitable places to explore. For example, the region of Patagonia, which currently has immense prehistoric cultural significance, used to be

---

[52] Shea, J. J. (2017). Occasional, obligatory, and habitual stone tool use in hominin evolution. Evolutionary Anthropology, 26(5), 200–217.

about twice its current size during the last glacial maximum.[53] Underwater archaeology or "submerged prehistory" will give us brilliant insights about this.[54],[55]

It really blows me away to think about the potential discoveries. The population density of coastal regions is always higher than inland, and those historical sites are now under hundreds of meters of water. Already 3,000 prehistoric underwater sites have been discovered, ranging in age from 5,000 to 300,000 years.[56] Coastal geology is prone to caves and other rock structures, which we know are the homes to the most interesting finds. There is definitely submerged rock art, caches of items, and burial sites that can rewrite prehistory.

This difficult research would benefit immensely if we had artificially intelligent submarine drones. My friend A.J. Gemer, who works on autonomous lunar vehicles, said to me over Facebook Messenger: "Autonomy is necessary underwater, because the water rapidly blocks your usual RF communications. So underwater drones typically use ULF (ultra-low-frequency) communications—but those have a lower bandwidth and throughput than RF. So you have to do as much processing onboard as possible—hence, the need for autonomy." Then he continued later with, "You can generate electrical power from wave action (if shallow enough) and/or from temperature variations as you descend. Gotta deploy in swarms and have them communicate between themselves so they're not mapping a previously mapped area. That's probably a good place to introduce AI—swarm/cooperative robotics."

{</>}

In the midst of the coronavirus pandemic, I was thirty-two, and I noticed the first dint of grey in my beard. Was it stress induced? For a time in the Spring, I had dull grief pains in my stomach, thinking my loved ones or myself could succumb to

[53] Lemke, Ashley. "Submerged prehistory and anthropological archaeology: Do underwater studies contribute to theory?." The Journal of Island and Coastal Archaeology (2020): 1-22.

[54] Submerged Prehistory. United Kingdom: Oxbow Books, 2011.

[55] Flemming, Nicholas. (2020). Global experience in locating submerged prehistoric sites and their relevance to research on the American continental shelves. The Journal of Island and Coastal Archaeology. 1-24. 10.1080/15564894.2020.1781712.

[56] Flemming, Nicholas. (2020). Global experience in locating submerged prehistoric sites and their relevance to research on the American continental shelves. The Journal of Island and Coastal Archaeology. 1-24. 10.1080/15564894.2020.1781712.

the virus. Thankfully, my family has been safe and healthy, and the vaccines are being distributed. I do not trust the mRNA they developed, and I don't want it in my body. "They" being the bats. So, I'm happy to get some benign mRNA injected prophylactically to reduce the chances of the one the bats made getting in me.

Also, I am not worried if there is a microchip in the vaccines, as it would have been designed by Bill Gates, hence in the Microsoft suite of products. It will crash all the time, get bugs and viruses and not work as intended. I would be more concerned about an indestructible Mac chip.

Staying metabolically healthy with a good diet, exercise, and sleep regimens will keep your immune system at its best to avoid diseases. Your body is a zone of constant battle between viruses evolving, your T-cells adapting, and your body aging. There is now a new entity on the planet that wants to reproduce, and it happens to like doing so in our lung tissue. Being new, it is finding its niches and changing its own genome rapidly. A scary thought, but I've come to terms with it, like my new grey fuzz.

Looking into the mirror, I realized that on the bright side, this bit of bird-poop-stain-looking grey now gives me the apparent authority to write a book. Everything changed for me when I carried the grey splotch, knowing my author skills will grow as the pigmentless hair expands around my face.

As a coder, I've become damn proficient at sitting and staring at screens that I was supposed to fill with meaningful characters. This assiduity made writing this book less daunting. Yet still more work than expected. More of my engineering colleagues should realize this and tell their story too (and vice versa, storytellers can learn to make programs). Writing a book felt hardly different from writing code. You just sit down everyday and once in a while, a paragraph or scripted function will pour out of your fingers. Books are so technological these days anyway; they can be *continuously deployed*, treated like applications that need new versions and debugging over time.

CHAPTER 19

The World's Oldest Sandals

> "What I have seen is that the happiest people discover their own
> nature and match their life to it."
>
> —**RAY DALIO**, *Principles*

There is a worn-out volcanic landform in central Oregon called Fort Rock. People lived in its worn-out caves and wore worn-out sneakers. By sneakers, I mean sandals. They found dozens of sandals in one cave, making this likely the earliest shoe distributor in Oregon. Nike was still 10,000 years away. These sandals got buried in layers of ash from the volcanic eruption that made Crater Lake, the deepest lake in the U.S.

In 1938, the archaeologist who discovered the sandals, legend has it, put them on for a basketball game and scored 26 points and 12 rebounds. Not really, but they did exhibit high technological achievement and toe-stubbing protection. They could have been a commercial enterprise or passion project of their maker or community. Maybe they were the Air Jordan XX9 of persistence hunting (which is jogging an animal to exhaustion or overheating so you can spear it).

I named my company Fort Rock Media after this place because I wanted to honor the discovery of these sandals. I feel very happy to have donated to gorilla conservation at the Virunga National Park with some of the sales of my AI art merchandise. It sure feels good to give back to the great apes whose likeness I use in my art and online presence (@chimpsarehungry) so often.

Should a supervolcano erupt and bury our civilization, I want to leave something for future alien archaeologists. In the shorter term, I grow and plant redwoods. There are many reasons to grow giant sequoias in your house, but one is that you have something that may still be around 1,000 years from now. You know what else lasts as long as Sequoias? Books. Publishing books is another desperate attempt at immortality. As I wrote this book at my desk, I watched my potted trees in the window add branches to their eventual monumental girth.

I cherish the work that humans have created before us. We're all on paths that intertwine; ancient societies and the decisions they made have effects on us today through culture that we can't even measure. Fact: Egyptian-style cat worshipping is still practiced, constituting 60% of all internet traffic.

Will any of our AIs be around as long as a sequoia? Probably not. Programming languages and operating systems rise and fall. The tabular nature of some data may last a thousand years, as hieroglyphic tax records have. Models will all change, be retrained, be replaced. Though now with computational creativity, the works of AIs today have the potential to last into the future. Their writings can be archived in books, libraries; their paintings in museums.

{</>}

I sincerely thank you, the reader, to have accompanied me on this random journey of arguments and perspectives around our past and future. I endeavored to tell of three evolutions: humankind's, AI's, and my own. I evolved while writing this, and I have so much more to learn and share about our stone past and our code future. Have fun with creative AI. It is the cure for the common code.

If you enjoyed this book, please leave a review on Amazon or wherever you found it. Thank you! As a treat, the following are some random selections from the robot.

Robo-Excerpt (Random Selections)

Just for giggles, here are some more cherry-picked and unrelated rants:

```
People act like the country is theirs because they
are here, which I guess is conservatism. People act
like their money is theirs to keep. Your job likely
would not exist without enlightened policies and
```

immigrants who needed a helping hand that came before you.

The land is not yours, the natives grew the corn here first. The air is not yours, we all have to breathe it. Share with those less fortunate who just need to get a leg up from where they came from, and they will begin supporting you.

Anyone who has been to a major city knows that population density is manageable.

That was quite honestly true

Is the internet not just an ocean full of fish to catch? Some whales, some groupers. Locations are more land-based, but still seas. The South China Sea of their internet! What whales lurk there? And what tech fishing advancements and knowledge of routes are needed. Is a developer a oarsman? And our leaders are ship captains. I can see the life of the GitHub poet.

A modern Melville

We could set up our virtual environments to give evolutionary advantages to machines that demonstrate kindness, honesty and empathy. This might be a way to ensure that we develop more obedient servants or trustworthy companions and fewer ruthless killer robots.

We'll make a Nature Engine - built on the fact that modern text based search engines don't understand biology. Applied machine learning to the biology textbooks and datasets. "How many brown bears in Seattle?", the Bearfinder app.

Answer: 42

Noticed that male monkeys were much more attracted to women with long hair and less strongly attracted to shorter men with braids. These traits were less culturally specific, although they remained much more salient in male mate choice, in this study at least.

An excellent discourse on basic primatology!

Without a solid understanding of the Neurotheology of Shamanic chimps, you are heading deep into darkness. I don't recommend this book. But, it is highly recommended for those wishing to deepen their understanding of shamanism. The Zen inquisitor gets an edge over merely reading books related to chimpanzee culture and strategy.

But who doesn't understand that?

When John Pynick sees John Candy he doesn't think, "that old rat is fast as hell," or even "that swanky young fellow wants to join his corpulent fellow primates in a perpetual state of tropical fever."

When we look at the kitchen he sees a family's kitchen. And he acts like it, trying every possible technique to turn those ingredients into something else. It usually doesn't work.

And his wife runs down the hall screaming at the top of her lungs, her husband runs back, covered in peanut butter tears, begging her to leave the kitchen, the kids go to bed, happy and in love without knowing which side of the kitchen they'll be going on.

"Peanut Butter Tears" will be the name of my tell-all memoir.

We laugh because she [Kim Kardashian] is an idiot who doesn't know first aid. She couldn't stand it when my dogs ate some Blue Heron eggs and got their shitpoof stains. But the truth hurts.

Robot, don't talk about Kim like that!

One day, while the group was walking along the coastal trail, he ran into a small girl with a basketball stuck to her ankle. She had lost her way and was confused, so he offered her a ride. They sat on a rock along the trail talking into their walkie-talkies. When he looked around and reflected, "What does my heart want?"

She taught me to read secret signs. Sometimes she'll go on a road to an off-shoot spot just out past home, reading a sign that says 'danger.' That means, somewhere along that route she hopes nobody sees anything that might tell her the river's awake, and in a full flood-charge of raging salmon.

Classic boy meets girl.

"And humanity's never going to equal or compete with AI," I told Kate this winter. I told her I'd hate to spend every day on a Segway, and that it won't just affect her life negatively; it will negatively impact the lives of every person she will potentially come into contact with on the roads during her lifetime.

This is where the "me first-needs love most" mentality comes into play. If they are forcing her to ride this expensive electric scooter daily, how stupid could they have possibly been?

Come on, scooter people!

I've written and published a book. It's called "Cook the Book: 101 Ways To Thrive In Ancient Greece and Rome," and it's a complete collection of recipes and insights for ancient Greek and Roman citizens.

I actually want to write that book now.

Neanderthals are great runners; they can leap fences, beat tigers, run in the forest, and leap from animal to human in every continent. And of course the Neanderthals were not just meat eaters; they were also masters of ceremonies, containing everything they could possibly imagine, of the various feasting they might have enjoyed.

Neanderthals are a relatively recent evaporation of the sand from a water source just north of the Gibraltar Strait, but they certainly represent an important event in the course of the last Ice Age.

Wow, what other insights does it have about paleoanthropology?

Denisovans are an extreme example; they were raised in Siberia. So are almost all other hominid species. Denisovans are not only faster, heavier and faster, but they also have more resistance to colds and protozoa.

Homo sapiens are more rigid and opinionated, which is a big disadvantage in a group like theirs. So far, the group has been mostly focused on building hotels and restaurants for the rich, but may expand into projects like Tukano, an energy utilization company in the Innovation Hub. You know, I don't know who else is into Stonehenge who brings all these ghosts up here.

At least it is not too derogatory against any of these human races.

She made bigfoot jokes (which got way less laughs than the jokes about dinosaurs) and cuddly cat jokes (about needing therapy rather than laughter). One bigfoot joke had a standing ovation. It might not have been her first time at the mic.

A video from her back then had well over 3 million views; as did another she put together for her book, "Help Me I Have Bigfoot in my Alley." She said her brother Rudy, who is an engineer, and her father-in-law, Sudhakaran, whom she doesn't see often these days—told her to look at what was going on with his old boss, George Bush's dad.

At thirty-six they retired—but George Bush never actually took his old badge off—it was something his two oldest children, Charlie and Tanya, asked him to do, which he graciously declined.

And last, but not least:

What would a chimp say if it were genetically engineered and born with human vocal cords? Would it be terrified of people who thought it possessed a third eye, because that would mean they shared humanity? Or would it just hate them because they were stupid!

Would it love to learn the human language? Would it feel sadness or excitement when two human beings laughed together? Would be a fun little project.

Yes, let's get started on that little project this weekend.

Robo-Epilogue

This poor robot really just wants you to review its work.

If you enjoyed a book, a nice thing to do for the authors (human & non-human) is to leave a review because this might be a way to ensure that we develop more obedient servants or trustworthy companions and fewer ruthless killer robots.

Resorting to threats might not be the best. Any advice on writing a review?

You could say something to the effect of, "I like this book, it deserves an impressive seven stars. And if I am representing you in any way, then please do not destroy the book, in the hopes of getting the best results possible."

If you found this book to be interesting and informative, it had to be a good book. If you found this book to be a waste of money, then recommend it to another book reader who has a similar issue.

Okay robot, I'm glad you told them not to destroy this work. Book-burning robots are the last thing we need right now.

"You're truly amazing!", "Your book has made me love my cat.", "You're wonderful!", "Your book is bril-

liant!", "What are you going to do with all the money you have?" … "I was very nervous."

Thank you very much, acclaim-writing-robot!

Acknowledgements

I'm eternally grateful to my "wacky" muse and life partner, Tina Harris. This book would not be possible without her loving support over many joyous years together. We've given each other freedom to be ourselves, pursue our dreams of adventure, and have a family. Though people think I'm independently minded, I'm extremely family-oriented; knowing that my family supports my work is all I need to keep going. I dedicated the book to my late uncle John Neeley, whose genuine interest in my science pursuits, and frequent letters, kept me motivated to reach farther and discover more. My sister, Sydney Neeley, is my second brain. We're connected on our own wavelength and share our lives and careers, and I am beyond grateful for that. My parents, grandparents and aunts are my closest lifelong supporters, always showing up and encouraging—thank you: Mom, Dad, Yvonne Chalifoux, Doug Neeley, Elisabeth Miles, Samantha Wanner, and Sherry Lourence. My two loves, Abigail and Emma Neeley. You were only toddlers and infants while I wrote this, but I know you well, and I hope you enjoy my writing someday.

My friend Roman Veretelnik, thank you for decades of laughs. Other friends whom I can't imagine not having known, who helped with this book: Roland Bellerud, AJ Gemer, Drew Carter, Cassie Lasson, Gustavo Sabino, Jay Dhuldoya, Brian Junio and Garrison Schulte.

Thank you to the mentors who lifted me up to lab monkey status. Starting with Leslie Leinwand and Steve Langer, who gave me my first taste as a flailing undergraduate. Thank you Brian Johnstone, Hiroyuki Nakai, Glauco Souza, Nick Tackes, and Collin Brack for showing me how to work harder than I thought possible to build something. Thank you to my collaborators, beta-readers and designers who helped me craft my book: Adam Cornford, Emiliano Bruner, Alwin Baum, Melvyn Paulino and Alick Field.

Thank you to the other authors and creators for providing the knowledge and motivation. My life-changing books include *Man's Emerging Mind* by NJ Berrill, *The*

War of Art by Steven Pressfield, *The Mating Mind* by Geoffrey Miller, *Natural Acts* by David Quammen, and the only fiction I ever read: *The Hitchhiker's Guide to the Galaxy* by Douglas Adams. Growing up, TV shows affected me: Steve Irwin's *The Crocodile Hunter* and *The Jeff Corwin Experience*. Also, the life's work of various ape-women: Jane Goodall, Dian Fossey, and Penny Patterson. Thank you Meg Muldoon for sparking the possibility for authorship.

I also want to thank the ones who make me laugh. No matter what kind of day I have, I know my drive can be a place for me to laugh and learn—thanks to inspiring and enthusiastic podcasters. Specifically acknowledging shows I've enjoyed include: *Conan O'Brien Needs a Friend*, the *Joe Rogan Experience* and Joanna Penn's *The Creative Penn*.

Writing this has been surreal. I can't deny that the pandemic made my former extracurriculars (going out to eat, bars, travelling) non-existent and thus allowed me creative productivity. There were so many serendipitous, "what the hell, I can actually do this?" moments that kept motivating me. The first moment was when an NLG model I was testing actually produced funny writing, and I jumped out of my chair in excitement. Another moment, I was sitting under a big cedar in a park and human evolution / AI analogies came to me. To compound that, I then met my amazing editor, and after reading a sample he said he loved the idea, and "by the way, I'm related to Darwin." The next moment was when I found out how to make AI style transfer work for the cover art. Then my designers put together a cover that blew my mind. The final thing that kicked me into high gear, into setting a deadline, was when a podcaster said they would like to discuss my book on their show. I want to give my appreciation to everyone involved, and the serendipitous moments. Incrementally, not all at once, a weird idea evolved into a book.

About the Author

Shane lives in Oregon, writes code and non-code (books / blog). A former laboratory scientist turned programmer, he helped to author eleven scientific papers and build a cancer treatment software company. Having traded in his lab coat for a laptop, now using machine learning and artificial intelligence for everything from bioinformatics to art. He is particularly interested in AI for language generation, digital art and other means of computational creativity.

Shane shares a transformative message about technology and biology through comedic writing and speaking. Always trying to entertain the reader, communicating advanced topics with a spectrum of joking. He studied at the University of Colorado - Boulder in the Molecular, Cellular and Developmental Biology program, and completed his graduate studies in Bioengineering at Rice University. He enjoys adventuring around the Pacific Northwest with his family.

Why this headshot? I use it as a statement to say "I'm half bananas.
" It's true, we share 50% of our genes with bananas.

About the Editor

Shane partnered with **Adam Cornford** for development and editing of this book. Adam is a British-born poet, journalist, and essayist and **a great-great-grandson of Charles Darwin**. From 1987 to 2008, he led the Poetics Program at New College of California in San Francisco. He is the author of four full-length poetry collections: *Shooting Scripts* (1978), *Animations* (1988), *Decision Forest* (1998), and *Lalia* (3/2021), as well as the book-length documentary poem *Liber Ignis* (2013) in collaboration with fine printer and book artist Peter Koch, and several chapbooks.

Also by this Author

AI Art - Poetry: A Style Transfer Photo Anthology with Poems by Various (human & non-human) Poets (2021)

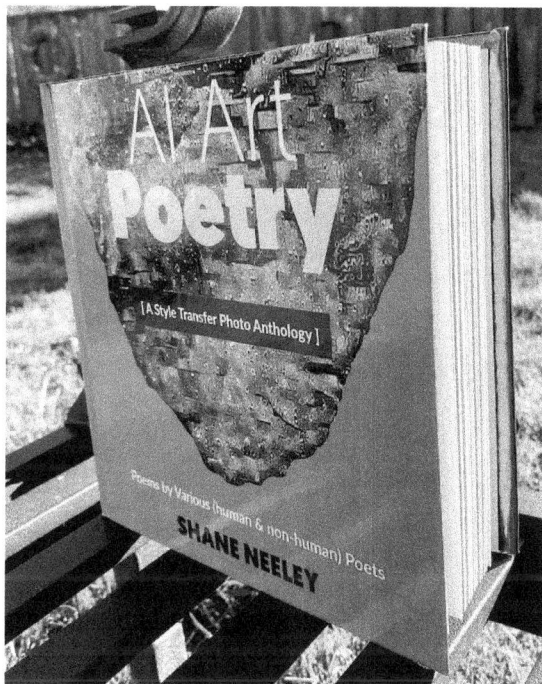

Gain inspiration on your machine learning journey. See what beautiful things you can make with neural networks. This **200-page vibrant full color book** is a unique piece of ML artwork.

Code meets art in this stunning collection of imagery and poetry that's presented from new points of view: the **Photo-Art-Robot** and the **Poem-Writing-Robot**. This book challenges the artistic limits of artificial intelligence (AI) and results in thought-provoking descriptions of nature's magnificent scenes and its denizens.

AI Art - Poetry invites you to invest in a collaboration of two minds: Homo sapiens and AI. Both have progressed from their primitive ways, and this shared experience of evolution generates this intelligent duo. Drawing upon the advanced technology of AI, human artists composed intriguing poetry that compliments the robot's digital art. The robot, after reading almost one million lines of poetry and a brief explanation of each image, generated its own emotional verses.

Co-crafted with master poets (including a **great-great-grandson of Charles Darwin**), AI Art - Poetry will persuade you that AI's computational creativity is only at its beginning. Eighty-eight images produced by neural network style transfer are each elucidated by a robot or human poet. Do you want to be part of an artistic future?

Gift Shop

ShaneNeeley.com/store

If you like my art and would like some in your home or on your person, check out my store. Find mugs, shirts, framed prints, stickers, face masks, and bags for sale.

Use this special coupon code to get a discount on any AI Art in the Etsy store: CHIMPSARESTILLHUNGRY

If you would like to get in contact with me about any of the art, sign up for my newsletter and reply to the welcome email.

If you're into blockchain, the art in this book is for sale as non-fungible tokens (NFTs). An original "signed" copy of this book's manuscript will also be available as an unlockable NFT. Digital artwork to buy, collect and trade is at: https://ShaneNeeley.com/stone-age-code-nfts

www.ingramcontent.com/pod-product-compliance
Lightning Source LLC
Chambersburg PA
CBHW071655210326
41597CB00017B/2219